From Galileo
to Einstein
Starting on a Journey in Science

From Galileo to Einstein

Starting on a Journey in Science

Tadayoshi Shioyama

Kyoto Institute of Technology, Japan

W World Scientific

NEW JERSEY · LONDON · SINGAPORE · BEIJING · SHANGHAI · HONG KONG · TAIPEI · CHENNAI · TOKYO

Published by

World Scientific Publishing Co. Pte. Ltd.

5 Toh Tuck Link, Singapore 596224

USA office: 27 Warren Street, Suite 401-402, Hackensack, NJ 07601

UK office: 57 Shelton Street, Covent Garden, London WC2H 9HE

Library of Congress Control Number: 2024943389

British Library Cataloguing-in-Publication Data
A catalogue record for this book is available from the British Library.

FROM GALILEO TO EINSTEIN
Starting on a Journey in Science

ISBN 978-981-98-0057-5 (hardcover)
ISBN 978-981-98-0058-2 (ebook for institutions)
ISBN 978-981-98-0059-9 (ebook for individuals)

For any available supplementary material, please visit
https://www.worldscientific.com/worldscibooks/10.1142/14034#t=suppl

Typeset by Stallion Press
Email: enquiries@stallionpress.com

Preface

Our current lives are a result of scientific evolution to which geniuses in science had contributed during preceding several centuries. This book describes the lives of great scientists from Galileo to Einstein who made remarkable discoveries and contribution to the progress of science. By focusing on their stories, the reader will understand that the common thread shared by them in their scientific journey was a genuine enthusiasm to scholarship.

The progress of science from Galileo to Einstein is surveyed as follows:

Galileo called the father of modern science, expressed natural phenomena with quantitatively measurable quantities, such as weight and length for the first time. Furthermore, on the basis of experiments, he explained the laws of physical phenomena in mathematical word. He discovered the law of falling bodies, etc.

Analyzing mathematically the precise data of astronomical surveys by Tycho Brahe, Kepler who was an astronomer and a mathematician, discovered Kepler's laws that explained the movement of celestial bodies, supporting the heliocentric theory proposed by Copernicus. It was however not understood why celestial bodies in the universe moved according to Kepler's laws. This problem was elucidated by Newton who inherited Galileo and Kepler and explained the motion of celestial bodies with the three laws of motion and the law of universal gravitation as described in *Principia*. The mechanics founded by Newton was called Newtonian mechanics that became dominant until the end of the 19th century.

Electromagnetic phenomenon was researched by Faraday who discovered the electromagnetic induction that transforms magnetism to electricity, Faraday's effect (magneto-optical effect) and so forth. Faraday, furthermore, proposed the concept of lines of force. Based on such experimental results by Faraday and the concept of Faraday's lines of force, Maxwell developed the concept of similarity between electromagnetic phenomenon and hydrostatics, to understand electromagnetic phenomenon. Then Maxwell succeeded to unify theoretically electromagnetic phenomena, and laid the foundation of electromagnetic theory. Newtonian mechanics and electromagnetic theory together constituted the two greatest theories in classical physics until the end of the 19th century.

However, at the end of the 19th century, two experimental results which could not be elucidated by classical physics, were observed.

The one was the experimental result on the relation between intensity and frequency of black body radiation, that was difficult to be resolved by classical physics. In 1900, Planck derived Planck's formula which strictly adhered to the experimental results of black body radiation. From physical interpretation of the formula, Planck discovered the concept of "quantum." In 1905, introducing the concept of quantum, Einstein succeeded in theoretically elucidating the phenomenon of photoelectric effect where an electron was emitted from the metal irradiated with light. This success confirmed the concept of quantum to be true, and then quantum mechanics based on the concept of quantum was founded as an epoch-making mechanics following Newtonian mechanics.

The other was the result of Michelson–Morley experiment on the measurement of the velocity of light in 1887, supporting relativity principle, that could not be explained by classical physics. According to the relativity principle, the velocity of light measured in any rectangular coordinate system moving at a constant velocity (called inertial frame) should be constant always. However, in classical physics, the velocity of light was different in different inertial frame, and therefore, the experimental result by Michelson–Morley could not be elucidated by classical physics. In order to resolve this contradiction, Einstein revised the concept of timespace on the basis of Lorentz transformation and founded relativistic theory supporting relativity principle. Thus, Einstein contributed to foundation of relativistic theory that constituted the two greatest theories in modern physics together with quantum mechanics.

This book describes progress in science from Galileo to Einstein. For the purpose of helping readers to understand this book, explanations and appendices for scientific knowledge and academic terms, are set at suitable positions.

The author has written this book with the hope that youthful readers worldwide will impress the lives of geniuses who carried out their end with a genuine enthusiasm to scholarship, and readers obtain the guides of their lives.

Acknowledgements

The author would like to show his appreciation to Professor K. Matsuda at Faculty of Law in Rikkyo University for providing pictures used in this book, of Newton's manuscript in the University of Cambridge and of Faraday's museum in London.

The author also thanks Professor K.K. Phua (Editor-in-chief), Ms Carmen Teo (Senior editor) and all members in the publisher for facilitating the publication of this book. He also appreciates other authors for the books referred to in this book.

Tadayoshi Shioyama
In Kyoto
May 2024

Contents

Chapter 1

Galileo Galilei

Galileo Galilei (1564–1642).

> Galileo expressed natural phenomena with quantitatively measurable quantities such as weight and length, and explained physical phenomena with mathematical words on the basis of experiments for the first time. Galileo was called the father of modern science.

1.1 Upbringing

1.1.1 *Galileo's birth*

On 15th February in 1564, Galileo was born in the neighborhood of Pisa, Italy, as the eldest among seven brothers and sisters. His parents were

Fig. 1.1.   Galileo's birth place (Pisa).

Vincenzo Galilei and Guilia Ammannati. His father born in Florence in 1520 was a teacher of music and also a fine lute player. His mother born in Pescia got married to Vincenzo in 1563, and lived in the neighborhood of Pisa.

His father Vincenzo studied phonics, quantitatively analyzing phonology. It seemed that his father's method influenced Galileo to express natural phenomena with quantitatively measurable quantities.

1.1.2 *The university of Pisa*

When Galileo was 10 years old, he received a private lesson by Jacopo Borghini in Florence where his family lived. When he was old enough to be educated in a monastery, his parent sent him to the Camaldolese Monastery at Vallombrasa which was situated on a magnificent forested hillside 33 km southeast of Florence. The Camaldolese Order combined the solitary life of the hermit with the strict life of the monk and soon the young Galileo was attracted to this life. He became a novice, intending to

Fig. 1.2. The Camaldolese monastery at Vallombrosa.

join the Order, however this did not please his father who had already decided that his eldest son should become a medical doctor, because there had been a distinguished physician in his family in the previous century. Vincenzo had Galileo return from Vallombrosa to Florence and give up the idea of joining the Camaldolese Order. However, Galileo did continue his schooling in Florence in a school run by the Camaldolese monks (http://www-history.mcs.st-andrews.ac.uk/Mathematicians/Galileo.html).

In 1581, Vincenzo sent Galileo back to Pisa to enroll for a medical degree at the University of Pisa. Galileo was however really interested in mathematics and natural philosophy.

Though the university had originally been established in Florence in 1343 as a Studium Generale, it was relocated to Pisa by the de' Medici in 1545. At that time, Pisa had been politically joined with its long-time rival Florence, thanks to it having been sold to the Duchy for 206,000 florins. Under the authority of the de' Medici family, the two cities became closely intertwined, and Duke Cosimo I de' Medici reopened the University of Pisa officially in 1545. His patronage greatly improved the standing of the school, which became one of the most prestigious in Europe (History, 2019).

In the year 1582–83, Ostillo Ricci who was the mathematician of the Tuscan Court, taught a course on Euclid's Elements at the University of Pisa, which Galileo attended. During the summer of 1583 Galileo was back in Florence with his family, and he invited Ricci (also in Florence where the Tuscan court spent the summer and autumn) to his home to meet his father. Ritti tried to persuade Vincenzo to allow his son to study mathematics because this was where his interests laid. Vincenzo did certainly not like the idea and resisted strongly, however eventually he gave way a little and Galileo was able to study the works of Euclid and Archimedes. He was still officially enrolled as a medical student at Pisa, but eventually, by 1585 he gave up this course and left without completing his degree.

1.2 Professor of Mathematics

After leaving halfway the University of Pisa, Galileo provided a private lesson of mathematics in Florence. In 1585–86, he taught mathematics at the Siena school. In the same year, he wrote his first paper "The little balance; La Balancitta," where he expressed the Archimedes' method finding specific gravity with a little balance. Afterward he went to Rome for the purpose of visiting Clavius who was the Professor of Mathematics in Jesuit College Romano. Although Galileo gave a favorable impression to Clavius, he failed in acquiring the position teaching mathematics in the University of Bologna.

Explanation 1.1 Archimedes' principle

This principle insists that a substance dipped in fluid receives buoyancy equivalent to weight of water with the same volume as the substance as Fig. 1.3.

Hiero II of Syracuse gave gold to a goldsmith, and had him fabricate a crown which would be offered for shrine. After the crown was delivered, Hiero II had a doubt whether the artisan stole a piece of gold and deceived by mixing silver weight equivalent to the stolen gold weight with the rest of gold. If the artisan made such deception, the volume of the crown should become larger than the volume of former gold. However, identifying the volume of the crown requested that the crown was melt and become a cube in order to simply measure the volume. Hiero II hoped the method that could measure the volume of the crown without destroying the crown. Hence he sent for Archimedes and asked the problem mentioned above.

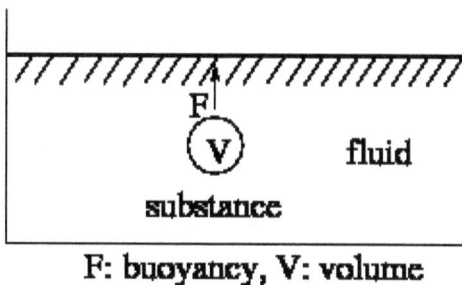

F: buoyancy, V: volume

Fig. 1.3. Archimedes' principle.

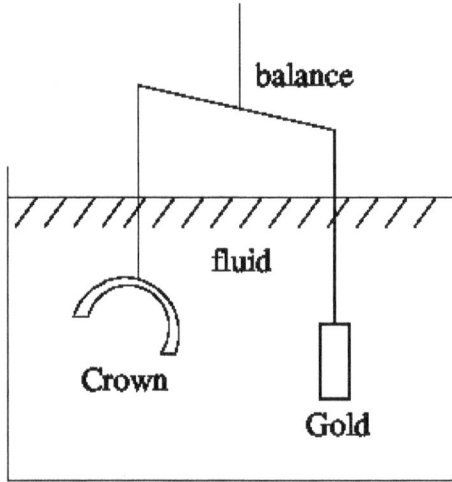

Fig. 1.4.   Decision of difference in specific gravity using buoyancy.

Archimedes solved the problem using principle of hydrostatics called "Archimedes' Principle". He hanged the crown at one end of balance beam and hanged the gold balancing with the crown at the other end. Then he dipped both the substances hanged at ends of balance beam. If a piece of gold was replaced with equivalent weight of silver, the balance beam should lean so as to raise the crown side of beam upward, because the specific gravity of the crown mixed with silver decreased and the volume increased, increasing the buoyancy of the crown. He confirmed whether the gold crown was mixed with silver, using this method.

After Galileo left Rome, he corresponded with Guidobaldo del Monte and Clavius for a long time since 1588. In 1589, since Fantoni left the position of Professor of Mathematics in the University of Pisa, Galileo was appointed as the Professor, because not only he was strongly recommended by Clavius, but also he gained a good reputation through his excellent lectures. He began to write the manuscript on the movement of a body.

In 1591, his father Vincenzo passed away. Since Galileo was the eldest son, he should support economically the rest of the family. Especially he should provide the dowries for the two daughters. He asked the

Fig. 1.5.   The University of Padua.

position with higher wages. On the strong recommendation of Guidobaldo del Monte, he was appointed as Professor of Mathematics in the second old University of Padua (Padova; Italian), earning three times wages.

1.3 The Law of Pendulum and the Law of Falling Bodies

When Galileo was looking at a chandelier's swing in the Cathedral of Pisa by using his own pulse rate, he discovered that the periodic time of pendulum is constant independently of the amplitude of the swing in a case of the same length of the thread.

He also discovered the law of falling bodies, where one fact was that the time during free falling is independent of the weight of the body, and the other fact was that the falling distance of the body is proportional to the square of time during free falling. In order to prove the law of falling bodies it seemed that he released two different weight of bodies at the top of the Leaning Tower of Pisa, and indicated the simultaneous arrival to the ground in both the bodies.

In fact, he used the slope on experimenting, because in the case of the slope it is convenient to measure the relation between the time and the

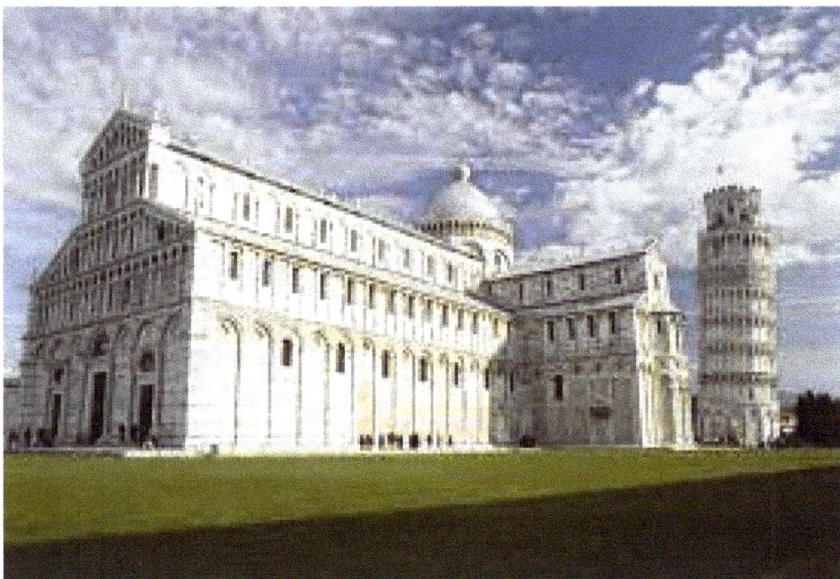

Fig. 1.6.   Cathedral of Pisa and Leaning Tower.

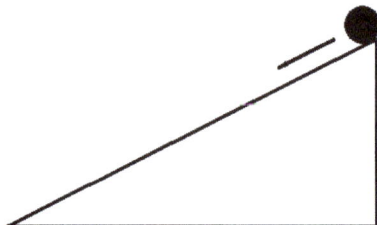

Fig. 1.7.   Experiment of falling body using slope.

falling distance rather than in the case of vertical falling motion. In 1604, he discovered the law of pendulum and the law of falling bodies. In 1608, he indicated the proportionarity of the velocity to the free falling time, and discovered that the moving path of a cannon was a parabola.

Explanation 1.2 Pendulum law

When as Fig. 1.8 weight set at end of light thread without elasticity is swung in a vertical plane, the period of pendulum is proportional to the

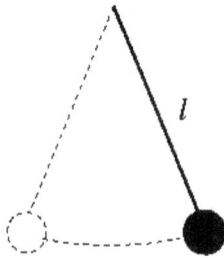

Fig. 1.8. Pendulum.

square root of thread's length. If the length of thread is constant, then period is constant independently of amplitude of swing. Pendulum law is applied to pendulum clock and metronome. By changing the position of weight, period is adjusted.

Explanation 1.3 Law of falling bodies

As Fig. 1.9 when an object with zero velocity is released, the falling by the gravity of the Earth is called free fall. Then, the velocity of the falling body is proportional to the time from the released time, and the distance in falling is proportional to the square of the time. From experiment, Galileo founded that the velocity and the distance of falling body were independent of weight. From this fact, it follows that the elapse from the time of release at some height to the time of reaching to the ground, is regardless of the weight of body, and the heavy body reaches to the ground at the same time as the light body.

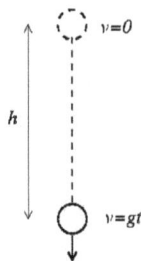

Fig. 1.9. Free falling.

Fig. 1.10.   Maria Celeste (portrait made in 17th century).

1.4 Marriage and Children

Galileo frequently visited Venezia where he met Marina Gamba 6 years younger than him and began a long term relationship with her. In 1600 the first daughter Virginia Galileo, in 1601 the second daughter Livia, and in 1606 their son Vincenzo were born. Two daughters enrolled in St. Matteo Monastery in their infancy. In 1616 Virginia became the nun called Maria Celeste. It seems that the nun Maria Celeste corresponded with her father. 124 affectionate letters for him were found after Galileo passed away.

1.5 Astronomy

At the beginning of 1609, Hans Lipperhey, an optician in the Netherlands, invented the telescope (Explanation 3.1). One story goes that when he happened to see the steeple of a church far away through a concave (凹) lens as the eyepiece and a convex (凸) lens as the object piece, to his surprise, the steeple appeared larger. This was a trigger for him to invent the telescope. Galileo who heard the story, polished the surface of a lens to a desirable curvature and succeeded in making a telescope with a magnification of 30 with a combination of 凹 and 凸 lenses and presented to the Governor of Venezia.

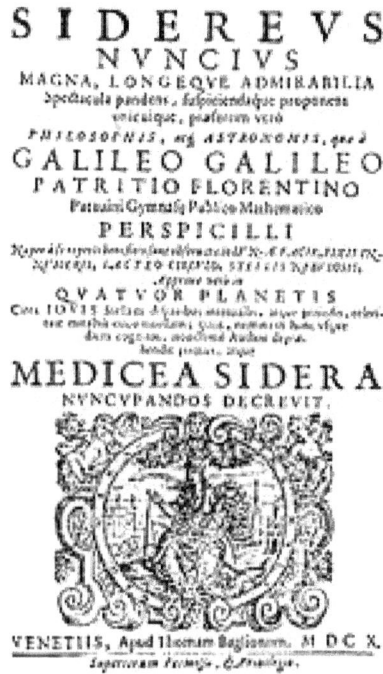

Fig. 1.11.   Report of stars.

On the other hand, Kepler fabricated a telescope with two ⬦ lenses. This type of telescope widened the range of view but had the problem of chromatic and spherical aberrations.

In the summer of 1609, Galileo identified a crater on the surface of the Moon with his own telescope. The following year, he discovered four satellites of Jupiter and published "Sidereus Nuncius (Report of Stars)" and his name came to be known across Europe (Galilei *et al.*, 1976).

On observing the Venus, Galileo discovered that not only the Venus repeated phases as the Moon but also varied the size. In the Putolemaic geocentric system, since the Venus passed through one side of the Sun it should be seen always as a crescent, contradicting the observation of Galileo. The observation of the Venus by Galileo was the evidence of that the Venus moved around the Sun. Furthermore, he observed the sunspots which were understood to exist on the surface of the Sun by Galileo. This posed a problem that even the Sun was never completely invariant.

Fig. 1.12.   The Grand Duke of Tsucany, Cosimo II (1590–1621); High relief of carved stone, gold and brilliant (high relief) 1617–24 Florence, Museo degli Argenti, cat 486.

In September 1610 he left Padua for Florence. He devoted all his time to research. He became the Mathematician and Natural Philosopher for the Grand Duke of Tuscany, Cosimo II de Medici (Sudget, 1981:41). Paying homage to the Grand Duke of Tuscany, Galileo named the satellite of the Jupiter as "Planet of Medici." In 1610, Galileo travelled for Rome, bringing his telescope. All people including astronomers in Collegio dei Societas Iesu (Jesuit) believed that the discovery by Galileo was the truth. He became a member of Accademia dei Lincei which was the first scientific association in the world. Accademia dei Lincei laid confidence to observation and experiments rather than philosophy and authorizing.

Appendix 1.1 Accademia dei Lincei

Galileo's visit to Rome in 1611 was a travel of triumph. Galileo was invited to Collegio Romano in celebration of "Sidereus Nuncius." In Accademia dei Lincei, a banquet was held for him and the new instrument was called "telescope." During his stay in Rome, he was recommended as a member of Accademia dei Lincei. His discoveries including the sunspot in Astronomy were shown to people in Rome. Galileo was allowed to lecture The Pope Paurus IV. All astronomers in Collegio Romano believed the discoveries by Galileo, excepting that Clavius doubted about mountains on the Moon and considered them to be hallucination. In 1613, the letter of Galilei concerning the Sunspot was printed in Rome by Accademia dei Lincei. The appendix of the letter included estimation on the positions of satellites of the Jupiter for several weeks from the planned publishing day, which was precise, and two eclipses were correctly estimated among three eclipses.

1.6 Satellites of the Jupiter

The accuracy of heliocentric annual orbital motion of the Earth, was confirmed by estimating eclipses of satellites of the Jupiter. He however showed the confirmation only in the appendix of the letter concerning the sunspot published in Rome in 1613.

Through the process improving the movement of satellites of the Jupiter in 1611–12, Galileo confronted with the fact that the Earth moved around the Sun.

Until October 1611, Galileo's calculation on the movement of satellites was geocentric, never heliocentric. From the end of 1611, he placed the center of orbital movement of the Jupiter to the Sun, never the Earth. Then, the positions of satellites of the Jupiter observed by Galileo coincided with their positions calculated on the basis of the observed time. His table of the satellites was sufficiently satisfactory. In Table 1.1, the orbital radii of four satellites of the Jupiter decided by Galileo in 1612 are compared with the corresponding modern values where the apparent radius of the Jupiter is used as the unit in measurement (Drake, 1993:185).

Table 1.1. Orbital radius of satellite of the Jupiter.

Satellite	I	II	III	IV
Galileo	5.5R	9.0R	14.0R	24.0R
modern	5.9r	9.4r	15.0r	26.4r

Galileo observed the eclipse of satellite of the Jupiter. He understood that an observer positioned at the Sun which is the center of orbital movement of the Jupiter, could not observe a satellite in case where the Jupiter existed between the Sun and the satellite. Hence he confirmed that the planet the Jupiter moved around the Sun as heliocentric theory insisted.

Since the discovery of the American Continent, the voyager requested the equipment precisely measuring the longitude of position of a ship on the sea. Although Tuscany was not the oceanic country, the Grand Duke of Tuscany, the employer of Galileo was interested in the equipment. Since Spain was the oceanic nation, the Tuscan ambassador in Madrid provided the equipment by Galileo to the Spain Government on September 1612. The equipment was fabricated on the basis of the discovery of eclipse of satellite of the Jupiter.

The longitude is the time difference between two places. It was conventionally measured by observing the eclipse of the Moon which did not so frequently occur. Hence Galileo prepared the table of position of satellites of the Jupiter. By observing the position of the satellite, a sailor could find the Florence time corresponding to the position using Galileo's table of the satellite. Then, the sailor obtained the longitude corresponding to the difference between the Florence time and the observing time on the sea. In fact, his table of satellites and related references were published after the other people had researched Galileo's table.

Appendix 1.2 Jovilase

From March 1612, Galileo devised an equipment called "Jovilase" which could measure the distance between satellite and the Jupiter. He described the strict scale drawing of 5 circles with the same center on a thick paper. The most inner circle was corresponding to the Jupiter, and the rest circles were corresponding to four satellites, as shown in Fig. 1.13. The horizontal diameter of Fig. 1.13 was divided into 24 sections using the radius of the Jupiter as the unit. At the most outer circle corresponding to the satellite IV, the angle number were written every 10 degrees counter clockwise from the far point (the top). A thread was passing through a hole at the center of the circle. At every section end in the horizontal diameter,

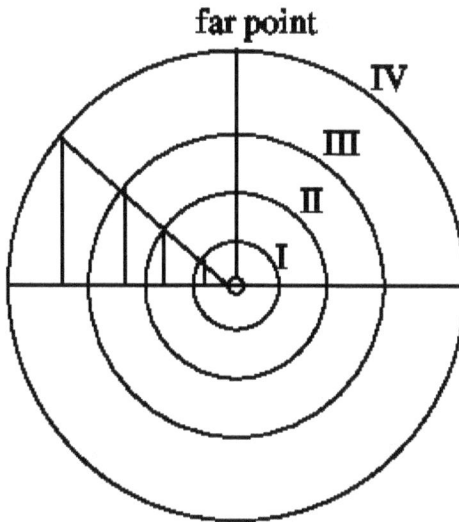

Fig. 1.13.   Jovilase.

perpendicular line was described, and at every four sections, number from 1 to 6 was described.

On the basis of the table of satellite, Galileo calculated the angle from the top of the time corresponding to requested position of satellite, then the thread was set along a line with the angle, and obtained the cross point with the most outer circle, where the perpendicular line was described, and found the cross point with the horizontal diameter, obtaining the distance of satellite using the radius of the Jupiter as a unit. The estimated distance by Galileo was precise. In 1617, Galileo furthermore devised the new Jovilase using brass instead of a thick paper in order to use it on the sea. The new Jovilase is preserved in Museo di Storia della Scienza.

1.7 Dialogue Concerning the Two Systems

In 1632, in Florence Galileo published "Dialogue Concerning the Two Chief Systems of the World — Putolemaic and Copernican; Dialogo sopra

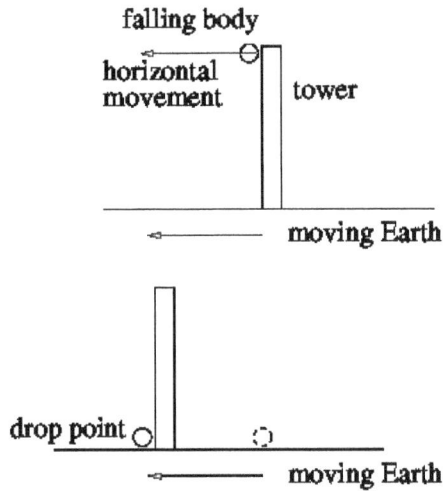

Fig. 1.14.   Free falling of body.

i due massimi sistemi del mondo." This book was the dialogue concerning the two systems where friends vivid discussed each other the problem whether they agreed with or opposed to each system.

Galileo described this book on the basis of the experimental result and the theoretical discussion. Since this book was written in Italian, never in conventional Latin, it could be read by even erudite people who did not take holy orders, being sold out several months later.

1.7.1 *The law of inertia*

Galileo used the law of inertia to explain the heliocentric theory. When one person insisted on the geocentric theory and said that if the Earth rotated then when a body fell from above, it should fall toward the west, Galileo explained that a falling body felt gravity in a vertical direction but no force in a horizontal direction and because the body falls with a horizontal velocity due to the rotation of the Earth, following the law of inertia, the body will fall straight to the ground below and not toward the west. It was however Newton, and not Galileo, who actually formulated the law of inertia.

1.8 Religious Trial

Although the new discovery by a telescope disturbed Astronomers, natural philosophers could be independent of it, because they considered that the system of celestial system was a mathematical fiction. It was the first confliction between Galileo and natural philosophers when Galileo disputed the problem concerning the principle of Aristoteles' Natural Philosophy.

1.8.1 *Hydrostatics*

The hydrostatics definitely had nothing to do with Astronomy, but the story was developed neglecting the general criterion. In fact, natural philosophers who opposed to Galileo insisting his hydrostatics, dragged Church into fighting against Galileo.

It seemed that floating of a body with higher specific gravity than water was in contradiction to the principle of Archimedes. In 1611, as soon as Galileo payed attention to the problem of a floating body, he found a satisfactory explanation concerning the problem, by examining the condition where a body with higher specific gravity than water could float on water.

In the case of such floating body, below the surface of water there was not only the body with volume V2 but also the air with volume V1 existing in the hollow of te body as shown in Fig. 1.15. That is, the buoyancy became the weight equivalent to the weight of water with the volume V1+V2.

Thus, Galileo understood that the body with specific gravity higher than water could float on water as a vacant kettle. Philosophers at that time

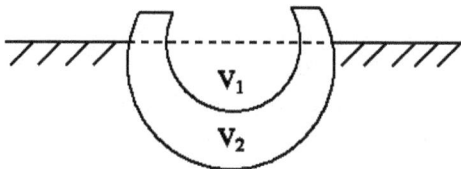

Fig. 1.15. Floating body; buoyancy = weight of water with volume V_1+V_2; V_1: volume of air below surface, V_2: volume of body below surface.

were ignorant of physics, and could not consider the explanation by Galileo. They merely followed Aristoteles who annotated the problem.

Galileo however knew the theory of Aristoteles in "On the Heavens, De Caelo," and understood it by reading the original.

Appendix 1.3 On the Heavens, De Caelo

This was the book of natural philosophy by Aristoteles, philosopher in old Greek (BC384–BC322), consisting of four volumes. He insisted that there was the fifth real element in addition to conventional four elements (fire, air, water and soil). The motion of the fifth element was continuously and forever complete. This element without weight was invariant eternally. The place where this element existed, was called "aether." The geocentric system was explained in this book.

Argument concerning hydrostatics in Florence became well-known. The Grand Duke of Tuscany recommended Galileo to write his insistence. Galileo said to the Grand Duke that if he would get a chance where he could discuss the problem with natural philosophers against Galileo before persons respected in the field of Philosophy, the Grand Duke could immediately judge whether Galileo suited the title of Mathematician and Natural Philosopher for the Grand Duke.

At the end of September 1611, Galileo got such the chance from the Grand Duke. At a luncheon in Tuscan Court with invited two cardinalises who happened to stay in Florence, Galileo discussed the scientific problem concerning the floating body with the Professor of Natural Philosophy in University of Pisa. Galileo showed the experiment supporting his insistence which satisfied the family of the Grand Duke. Although one of cardinalis supported natural philosopher insisting Aristoteles theory, the other of cardinalis Maffeo Valbelini who would become The Pope Urban VIII strongly supported Galileo.

After Galileo won the discussion in Tuscan Court, he wrote "A body in Water" from January to March 1612 in Le Selve, and published in Florence in March. This publication profoundly influenced the relation beween Galileo and natural philosophers who lost the discussion and afterward had theologians take sides with themselves. The group consisting of natural philosophers in Florence and Pisa united against Galileo.

In 1618, a Comet appeared. In the same year, a Jesuit priest Orazio Grassi in Collegio Romano published a book "On the Comet: De Tribus Cometis," where Grassi described that the Comet moved with the constant velocity on the sphere whose center was the Earth. In 1619, Galileo published "Discourse on the Comets" which was the controversy with Grassi. This argument yielded the separation of Galileo and Jesuit priests who had been friendly to Galileo.

As a result, several Jesuit priests reported to Church that "Dialogue Concerning the Two Chief Systems of the World" was very one-sided and partial.

It was the origin of the quarrel that natural philosophers following Aristoteles' theory were humiliated by Galileo on discussion concerning the floating body in Tuscan Court. The philosophers bearing a grudge against Galileo would like to take revenge on Galileo for this insult, and

Fig. 1.16.   Galileo on Religious Trial in Vatican; Joseph-Nicolas Robert-Fleury 1846 年 Ruble Museum.

Jesuit priests were dragged into the quarrel. Then the book "Dialogue Concerning the Two Chief Systems of the World" scientifically interpreting the Bible fell victim to a religious trial, yielding the situation where the Vatican could not stop suppressing the heliocentric theory.

The Pope put the book above mentioned on trial at the Inquisition. As a result, the book was imposed a ban on publishing. Galileo was summoned in 1633, and was on trial at the Inquisition being shown off torture tools.

Representatives of Pope Urban VIII pleaded with Galileo to simply recant his position on the matter and beg forgiveness, but he refused, time and time again (History, 2019).

Since he insisted the heliocentric theory, he was confined under the Priest at Siena. Galileo was condemned to lifelong imprisonment, but the sentence was carried out somewhat sympathetically and it amounted to house arrest rather than a prison sentence. He was able to live first with the Archbishop of Siena, then later to return to his home in Arcetri, near Florence, but had to spend the rest of his life watched over by officers from the inquisition.

1.9 Discourse and Mathematical Demonstrations Concerning the Two New Sciences

In 1638, Galileo published "Discourse and Mathematical Demonstrations Concerning the Two New Sciences; Discorsi e dimonstrazioni mathematiche intorno a due nouve Scienze, attendi alla meccanica ed ai movimenti locali" by Elsevier in the Netherlands. This book was the last published one by Galileo, and was the most scientifically important to him.

In this book, Galileo proved the law of falling body using mathematics. The law discovered by him is described in detail as follows:

$$1.\ s \propto t^2 \quad 2.\ v \propto t \quad 3.\ s = vt/2$$

where v implies the velocity of falling body, t the time measured from the static state, s the falling distance from the static state, and $\propto$ proportionality. It took several years of experiments, observation and mathematical research to derive these relations. In the days of Galileo, there was not yet mathematical methods for calculating differentiation and integration. Galileo did not recognize that the proportional coefficient to time of velocity was the acceleration g of gravity, which was elucidated by Newton.

In order to study the detail of motion, he fell a body along a slide, and used his own pulse as a clock. In 1609, he found the law that the velocity was proportional to time. Galileo could not distinguish mass from weight (a force influenced to mass under gravity). The problem had not been solved until Isaac Newton elucidated it.

In the days of Galileo, it was a very interesting problem how a shot body would move, among various movements, because the problem was especially important in cannon tactics and battle. But, the movement was very complicated. It was difficult to understand the movement because the law of falling body was not known before the days of Galileo.

In the case of cannon, on explosion of powder additional momentum is transmitted to a cannon. The effect of the momentum is instantaneous and decreases as time passes, ultimately disappearing. When a cannon is shot above, it does not move eternally upward, but at the top of the movement the momentum disappears and a cannon begins to free fall in accordance with the law of falling body.

On the other hand, it was found that the horizontal velocity of the cannon was preserved until the arrival to the ground as the result of experiments, according with the inertia. Galileo proved that two elements of the motion, consisting of one vertical element of free fall and the other horizontal element, were independent each other. The horizontal distance is the product of the constant velocity by the time, being proportional to the time. Since the vertical distance is proportional to the square of time, the

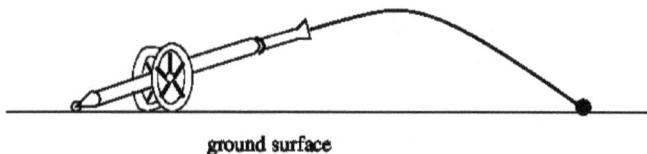

ground surface

Fig. 1.17. The moving path of cannon.

vertical distance is proportional to the square of horizontal distance, hence the relation between distances was shown as a parabola in the moving path of a cannon (Fig. 1.18).

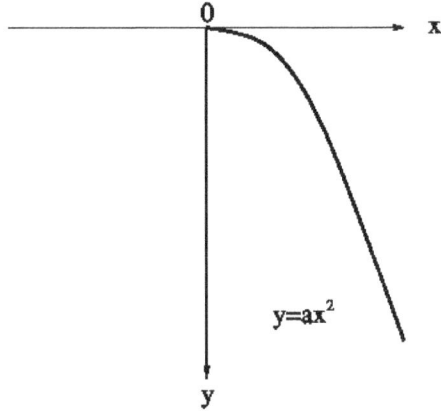

Fig. 1.18.   Parabola; x: horizontal distance, y: vertical distance.

Fig. 1.19.   Galileo later in Arcetri; Nicolo-(after)-Barabino.

1.10 Galileo Later

In 1634, Galileo lost his affectionate eldest daughter Maria Celeste suffering from a serious disease. Furthermore in 1637, he lost one's sight. Soon later, he lost both sights.

When Galileo published the book in 1638, his pupil Evangelista Torricelli orally wrote the manuscript. Torricelli discovered the Torricellian vacuum above the surface of the mercury in the tube (Fig. 1.20). The unit of pressure "Torr" was named for the purpose of praising his contribution.

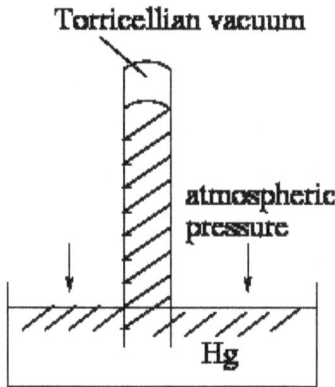

Fig. 1.20. Torricellian vacuum.

Fig. 1.21. Sculpture of Galileo; Museo di Storia della Scienza.

For several years of his later, he became frequently sickly. In November 1641, he developed kidney trouble, and became a bedridden old man. On 9th of January 1642, he passed away, buried in the family tomb in the Basilica of Santa Croce (Sugget & Oohashi, 1992).

1.11 The Revolutionary Points in Galileo's Research Work

The revolutionary points in Galileo's research work are as follows:

- Time was introduced as an elementary quantity of a physical phenomenon.
- Natural phenomenon that until then was expressed philosophically, was expressed with quantitatively measurable quantities such as weight and length.
- On the basis of experiments, he expressed the laws of physical phenomena in mathematical words for the first time. After him, this method of explanation became the most important method of scientists.

Fig. 1.22. Telescope of Galileo; Museo di Storia della Scienza.

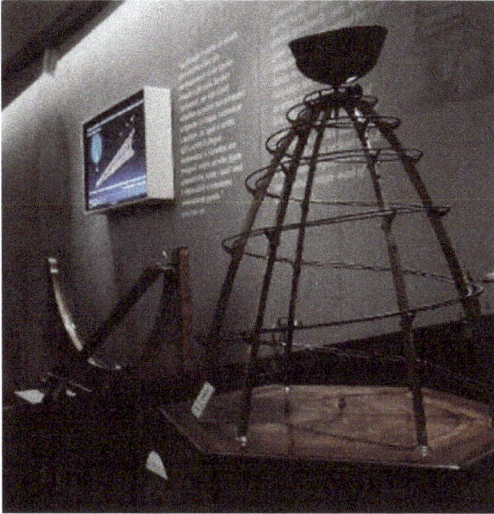

Fig. 1.23. Experimental equipment used for discovering the law of falling body; Museo di Storia della Scienza.

1.12 Since the Religious Trial

After deliberation concerning the religious reform at The Second Vatican Conference in 1962–5, at the plenary session in October 1992, the Pope John Paul II gave an address on behalf of the Catholic Church in which he admitted that errors had been made by the theological advisers in the case of Galileo.

360 years later, the Catholic Church withdrew the guilty sentence to Galileo.

Chapter 2

Johannes Kepler

Johannes Kepler (1571–1630).

> Johannes Kepler discovered "Kepler's laws" concerning the law of orbital motion of celestial bodies by analyzing the data of astronomical surveys by Tycho Brahe. He accomplished a revolutionary work in astronomy. His remarkable achievement was inherited by Newton who elucidated the motion of celestial bodies in universe.

2.1 Upbringing

2.1.1 *Kepler family*

Kepler family was once proud aristocratic family. In 1433 of one century before, Johannes' great-grandfather of great-grandfather was conferred a night title by Emperor Sigismund because his brave work in a battle was approved. Afterward the family had not devoted their life to Emperor, and changed their aristocratic class to artisan class, moving to a small quiet town Veil der Stadt. It was however still told that a great-grandfather and grandfather of Johannes were awarded medals by Emperor Kaar V.

Although Kepler family was never famous as before, Sebald who was the grandfather of Johannes, had served as a mayor at Weil der Stadt. He was a dictator, but was trusted by people of district because he advised adequately people.

Sebald was the leader of Kepler family and was more paternal than somebody else for Johannes.

2.1.2 *Johannes' birth*

On 27th December 1571, Johannes Kepler was born at the house of his grandfather Sebald in Weil der Stadt, and was the first son for his parents. His father Heinrich was the fourth son of Sebald and lived under paternal love. Heinrich was coarse person without education, and was frequently not in the house when Johannes was infant. Johannes wrote about his father as follows: "He yielded collapse of my family, and was blunt and quick to start a quarrel."

It seemed that Heinrich inherited the ancestor of Kepler family who had devoted their life to Holy Roman Emperor with the heart of brave

Fig. 2.1. Johannes' birthplace.

soldier. Since Heinrich who lived at a room too small for his family in the house of his father Sebald, felt suppressed, he went out of the house when his son was only 3 years old, and became a soldier in the employ of the Netherlands. When Johannes was a child, such a thing was repeated. Even when Heinrich got home, shortly soon he was fascinated by the battle field. In 1588, when Johannes was 16 years old, Heinrich went out the house and never got back home. It was uncertainly told that he worked as a Naval commander of Regno di Napoli, and on return from the battle he passed away in Ausburg.

Johannes' mother Katharina was a daughter of Melchior Guldenmann who was an innkeeper in a village Ertingen and a mayor of the village.

Johannes was raised by his mother, and was small and black. Both mother and son were full of curiosity. His mother was not educated, but interested in healer and a cure for disease by healer. When she grew old, she would be brought to trial because she was doubted to be a witch for such her hobby. She was certainly eccentric and not friendly with her neighbors. Johannes evaluated about his mother as follows: "She had a sharp tongue, was a wrong spirit, and was quick to start a quarrel." She had given birth to seven children but only four children could be grown up. Such infant death rate was usual in the 16th century.

In 1577, the Great Comet appeared and was observed over Europe. Johannes' mother Katherina went up a hill together with 5 years old Johannes in order to have her son see the magnificent view. The view of the Great Comet however did not give a strong impression to Johannes who had got smallpox when 4 years old and then had weak eyesight, never clearly seeing the view of Comet. But he did not forget that his mother had him see the Comet, as remembrance of his childhood.

2.1.3 *Empire free city Weil der Stadt*

Weil der Stadt where Johannes was born was a small city where about 200 people lived, and was also Empire free city. Free city was a city independent of surrounding duchy such as Wurtenberg in Holy Roman Empire which consisted of many duchies, territories of bishop and cities (Voelkel, 1999).

Holy Roman Empire included Germany, Austria, Bohemia, part of France and part of the Netherlands, and it was governed by Holy Roman Emperor Rudolf II at the Court in Praha, Bohemia. Although Wurtenberg duchy surrounded Weil der Stadt, which was free city, both Protestantism and Catholicism were permitted in Weil der Stadt.

2.2 Education

2.2.1 *Latin school*

Religious problem strongly influenced the education of Johannes. Among his brothers only Johannes received the education of university. When in 1577 5 years old Johannes started the first step of the education, the family had moved to Leonberg near Weil der Stadt. Since Leonberg was part of

Wurtenberg duchy, Johannes could use the splendid educational system which was constructed by the Duke for people.

Johannes at first attended German school and moved to Latin school soon, which was the school linked with the education of university. In Latin school, he learned reading and writing Latin which was the international language of scholarship, used over Europe. The pupils should speak only Latin each other. Kepler wrote all the letters and books in Latin even for German.

Kepler never went up through the educational system without a hitch, because in 1580–82, 8–10 years old Johannes who was small and weak, made hard farm work under his parents' order. Hence he was relieved when he returned to the school.

2.2.2 *Seminare Maulbronn und Blaubeuren*

Passing the national examination, in October 1584 Kepler was approved to enroll in the elementary seminary in Adelberg which was the first step among two steps to the enrolling in the university. Since Kepler had a good school record, 2 years later he enrolled in the advanced seminary, that is, Seminare Maulbronn und Blaubeuren whose building had been the monastery of Ordo Cisterciensis (English: Cistercians) once.

Kepler who was small and sickly, was vivid in the school, wrestling with difficult intellectual exercise in order to escape from the unpleasant remembrance during his childhood. He took delight in writing classic style poet, interested in metre. Also he learned the longest Psalms by heart.

Kepler was serious and faithful pupil. He had wrestled with religious problem since his childhood. He did not satisfy to learn what he was taught, and would like to confirm whether it was correct or not. Hence when he heard a sermon which denounced some denomination of Christianity, he judged himself whether it was in fact described in the Bible or not.

2.2.3 *The University of Tubingen*

As a result of Kepler's diligence in the school, in September 1588, he passed the Bachelor approval examination. Kepler who was still registered in Seminare Maulbronn, had been officially registered as a student in the University of Tubingen more 1 year before. That is, he completed the

Fig. 2.2. University of Tubingen.

study in the University of Tubingen at Seminare Maulbronn, and got the degree of Bachelor without attending the lectures in the university. He got the way to learn the theology in the university after getting the degree of Master.

In September of the next year, Ludwig der Fromme (Duke of Ludwig) listed five scholars for Tubinger Stift which included Kepler. Kepler would like to devote his life to the Duke of Wurtenberg, getting the scholarship, because everything necessary was provided. The Stift guaranteed to give a house until completing study for 2 years to get the degree of Master, and to be responsible for starting the Theology during 3 years. Kepler left for Tubingen. On 17th September 1589, he signed the register of the Stift as follows:

Johannes Kepler born on 27th December 1571.

At the time, he was 17 years old. After he would study in educational program of the university for 2 years, he could genuinely devote himself to the Theology. What Kepler was interested in and gave attention to, was the research field including Astronomy, Mathematics and the Theology.

2.2.4 *Mathematics and Astronomy*

Kepler learned Astronomy and Mathematics from Michael Maestlin in the University of Tubingen. Kepler profoundly respected Maestlin who taught Kepler about the recent Astronomy, that is, heliocentric system proposed by Astronomer Nicolaus Coperunics 50 years before. Maestlin was one among few persons who believed the heliocentric system to be true. He however taught the geocentric system, that is, Ptolemaic system to elementary students.

Appendix 2.1 Ptolemaic Astronomy

Ptolemaic Astronomy proposed by Claudius Ptolemaeus continued to be a chief Astronomy during 1500 years since the 2nd century. Ptolemaeus considered that the world was static at the center of universe, and was a sphere studded with fixed stars at the end of universe. He added mathematical theory concerning the motion of each planet to the fundamental framework mentioned above. The theory had been sufficient to estimate the motion of planet by some little improvement until the days of Kepler.

The universal theory of Ptolemaeus did not contradict with the theory of elements by Great Greek Philosopher Aristoteles in BC4 century. Aristoteles taught that celestial material was made of Aethel which was different from elements on the ground such as soil, air, fire and water, moved eternally on circular orbit, and was invariant.

In 1572, a dazzling nova appeared, and in 1572 the Great Comet appeared. These happenings seemed to contradict with the theory by Aristoteles which insisted for Aethel to be invariant. Tycho Brahe who precisely observed the Great Comet in Denmark, considered that Aethel did not exist.

Kepler considered that Copernican system included the religious wide significance, and that the universe was none other than the figure of the Creator.

Explanation 2.1 Copernicus

On 19th February 1473, Nicolaus Copernicus was born at Torrun in Royal Prussia. His father Nicolaus who was a wealthy merchant dealing copper, passed away when his son was 10 years old. His mother Barbara Watzenrode passed away by the time. Hence he and his brothers were raised by his maternal uncle Lucas Watzenrode who was the Canon, and later became the Bishop of Warmia in Royal Prussia. Since his uncle wished that he would become the Bishop, in 1491 Copernicus enrolled in the University of Krakow, where he learned Astronomy from famous Professor Albert Brudzewski who precisely calculated the orbit of the Moon for the first time.

After he ended his study in the University of Krakow in 1495 without getting the degree of Bachelor, he became the Canon of Warmia, obtaining guarantee of his living, and lived at Fromborg in Baltic coast. In 1496, he studied abroad in the University of Bologna where he learned the law of Canon. In the meanwhile, he met famous Astronomer Domenico Maria Novara da Frrara, and became his disciple. In 1500 he completed study in the University of Bolonga, and returned to Fromborg. After getting the

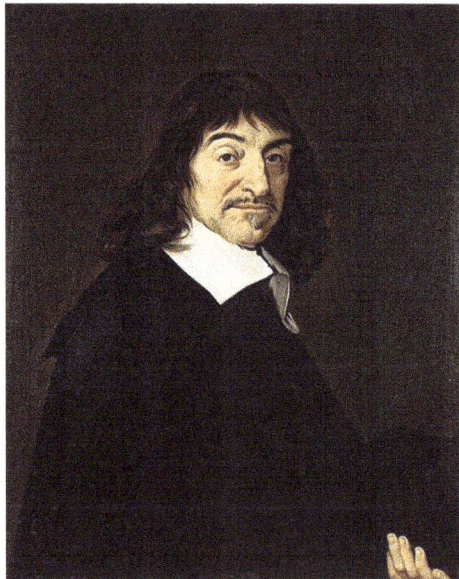

Fig. 2.3.   Nicolaus Copernicus (1473–1543) (portrait in 1580).

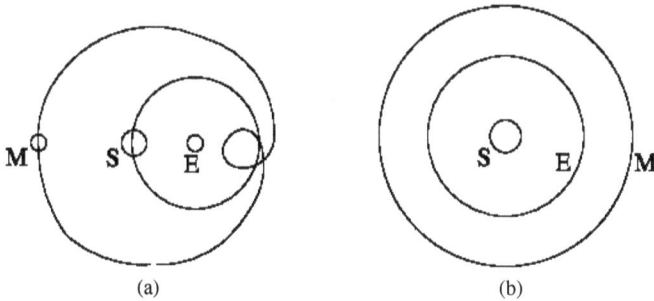

Fig. 2.4. Ptolemaic system (a) and Copernican system (b). S: the Sun, E: the Earth, M: the Mars.

permit of the cathedral Chapter in Warmia, in 1501 he studied Medicine in the University of Padua. He learned in Padua for 2 years, and in 1503 he got the degree of Doctor in the law of Canon. Then he returned to Warmia, and became the Canon. Afterward he did not leave Warmia.

He was busy with works not only as a clergyman but also as a doctor. On the other hand, he observed celestial bodies in his free time, and described his concept. It seemed that he got his concept of the heliocentric system in 1508–10.

In Ptolemaic system there was a reverse movement on the orbit of the Mars as Fig. 2.4. On the other hand, in Copernican system such reverse movement did not occur, and the celestial bodies' movement was simply understandable. He published in 1542 "De revolutionibus orbium coelestium (On the Revolutions of the Celestial Spheres)." He passed away on 24th May 1543.

2.2.5 *Change of course*

Kepler was studying Theology on schedule. On 11th August 1591, he completed learning advanced necessary subjects during 2 years, getting the degree of Master. Two months later, the board of trustee sent a letter for mayor and the municipal assembly of Weil der Stadt which requested to prolong the period giving scholarship for Kepler, writing as follows: "young Kepler had brilliant knowledge and could be expected to accomplish something especially marvelous."

At the beginning of 1594, the plan of Kepler however was changed. When he could complete studying Theology during 3 years, only few months later, he was coerced into stopping study of Theology. In the preceding year, Georg Stadius who had taught mathematics in a Protestant seminary at Graz in Steiermark, passed away. In November, members of assembly in Steiermarg requested that the University of Tubingen would recommend a prominent person as the successor. The department of Theology elected Kepler because he was with enthusiasm learning under Maestlin and outstanding.

It was to be separated from the chance devoting himself to Church rather than to go to far country Graz why Kepler was not satisfied. He did not wish to be appointed as a teacher of only mathematics, excepting a clergyman. On the other hand, he considered that a person was not sent to this world for only himself. As a result, he proposed compromise assuring the possibility that in future he could return to Church. The headmaster of Stift in Tubingen and school inspectors of Protestant school sent a letter to the Duke of Wurtenberg which requested that Kepler leaved Wurtenberg to be appointed as a teacher in Graz. The Duke of Wurtenberg agreed with the request on 5th March. On 13th March 1594, Kepler left the University of Tubingen for far Steiermarg.

2.3 Mathematical Teacher in Graz

2.3.1 *Religious coexistence*

When Kepler came to a new district, he first felt that the religious circumstance was very different from Wurtenberg which was a Protestant country, because in Steiermarg both of Catholicism and Protestantism awkwardly coexisted. According to the Religious Reconciliation in Augsburg, people should be Catholic as Hapsburg. But almost all leading aristocrats having territories in south Austria, were converted to protestant. Hence 20 years before, the Grand Duke Karl conceded that people could believe his own religion by the peace of Brug (1578).

The Protestant seminary in Graz was the first seat of power, and the officials of the seminary were important representatives of protestant society.

2.3.2 *Mathematical education by Kepler*

The boy's school, the seminary in Graz (the University of Graz in present) consisted of elementary and advanced steps. Kepler taught Mathematics including Astronomy at Philosophical class in the highest fourth form of advanced step. The number of pupils who attended the lecture of Kepler, however was few. School inspectors were aware that young Kepler was not wrong, but the subjects were not adequate. Hence from next year Kepler taught various subjects other than Mathematics and Astronomy, including rhetoric, poem of Vergilius (Publius Vergilius Maro) classic Roman poet, basic mathematics, history and ethics.

In addition to the work as a teacher, Kepler was also appointed as a Mathematician for the district whose work was to make almanac and to estimate occurrences in next year by astrology. Kepler had a complicated idea about astrology, because he disliked to nourish stupid superstition, and on the other hand he truly believed that occurrence such as Planets forming a line, exercised important influence on man and nature.

Fig. 2.5. University of Graz.

In 1595, Kepler estimated with astrology that there would be the severe cold weather, the aggression to south Austria by Turkey, and insurgence by farming population. Since this estimation was correct, Kepler got a bonus 20 florin gold. Thus, astrological estimation was an important income for Kepler.

2.3.3 *Theology and Science*

Although Kepler agreed without satisfaction that he taught mathematics in Graz, he decided that he would bring up his research on science to an appropriate philosophical level, when he was working as a teacher of Mathematics including Astronomy. Hence after he examined Copernican system, he became aware that there was some elements that were not sufficiently explained.

A feature of convincing explanation in Coperunian system was that the system contained planets' orbits in harmony. That is, in Copernican system the distance between the Sun and each planet was decided by their relation, placing each planet at the special position with distance from the Sun. The dimension of the Mercury's orbit was a third of the Earth's orbit, being about two third in case of the Venus, about 1.5 times in case of the Mars, 5 times in case of the Jupiter and 10 times in case of the Saturn.

By considering in detail the heliocentric system, Kepler was aware that Copernicus did not explain the fundamental reason why planets were thus placed, never in other way. Why were there six planets, and did the Creator select such form of heliocentric system, never other form of system?

2.3.4 *Mystery of the Universe*

A fundamental principle concerning a solution to the problem mentioned above occurred to him, when on 19th July 1595 he taught a equilateral triangle inscribed to a circle to which three vertexes of equilateral triangle contact. Kepler was aware that if drawing a circle inscribed to the triangle, the ratio between the dimension of larger circle and that of smaller circle is the same ratio between size of the Saturn's orbit and size of the Jupiter's orbit. If drawing a square inscribed to a small circle, and furthermore drawing a circle inscribed to this square, the ratio between these sizes might be the ratio between sizes of the Jupiter's orbit and the Mars' orbit. Kepler considered that all the planet's orbits were decided on the basis of geometrical principle and the Creator might found the Universe on the basis of geometry.

Kepler was aware that three dimensional geometry should be used instead of two dimensional geometry, therefore using sphere instead of circle and using regular polyhedron instead of polygon. It was known by old mathematicians Greek that there were only five regular polyhedrons such as tetrahedron, hexahedron (cube), octahedron, dodecahedron and icosahedron. Kepler wrote in preface of *Mysterium Cosmographicum* (*The Cosmographic Mystery*) as follows:

The celestial sphere of the Earth is a measure of the celestial sphere of planets. If drawing a dodecahedron circumscribed to the celestial sphere of the Earth, the sphere surrounding the dodecahedron is a celestial sphere of the Mars. If drawing a tetrahedron circumscribed to the celestial sphere of the Mars, the sphere surrounding the

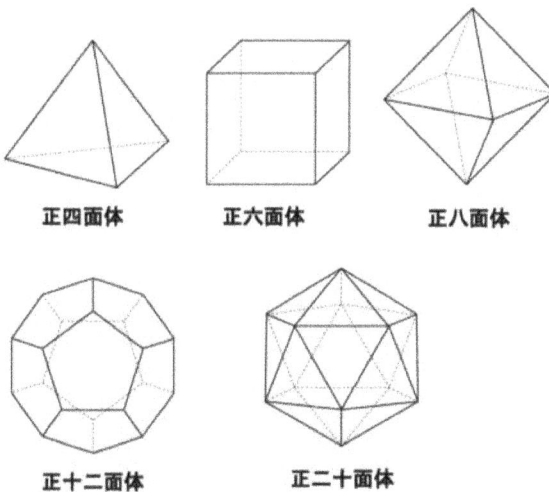

Fig. 2.6. Five regular polyhedrons. https://www.fy1203.com/2020/04/28/regular-polyhedron/.

tetrahedron is a celestial sphere of the Jupiter. If drawing a cube circumscribed to the celestial sphere of the Jupiter, the sphere surrounding the cube is the celestial sphere of the Saturn. On the other hand, if drawing a icosahedron inscribed to the celestial sphere of the Earth, the sphere inscribed to the icosahedron is the celestial sphere of the Venus. If drawing a octahedron inscribed to the celestial sphere of the Venus, the sphere inscribed to the octahedron is the celestial sphere of the Mercury.

Kepler understood the reason why there were only six planets. On 20th July 1595, he accomplished the discover that if there were six celestial spheres of planets, it was possible to inscribe five polyhedrons to these five exterior celestial spheres except the most interior celestial sphere of the Mercury. He wrote in the letter for his teacher Maestlin that he regarded his discover as an incredible miracle by the Creator. He decided to announce his discover by publishing a book, because he considered that such a book would be the proof of correctness of Copernican system and also of glory of the Creator. He considered that it was to elucidate the plan of the Creator building the Universe how he made the work as a mathematician become meaningful.

In January 1596, hearing that his grandfather was suffering from a serious disease, Kepler left Graz for visiting his grandfather. Old grandfather passed away by his arrival. At the time, he traveled to Tubingen for visiting Maestlin, and negotiated with a printer in order to publish a book, because in Graz there was not a printer who could print a book of complicated Astronomy, but in Tubingen a printer Grupenbach existing with capability for printing. In March 1597, he completed printing, but publishing year was 1596. The title was *Mysterium Cosmographicum* (*The Cosmographic Mystery*).

2.3.5 *Marriage*

On 27th April 1597, Kepler got married to Barbara Muller who was a daughter of Jobst Muller running a mill in south district of Graz. In the next year his son was born but passed away soon. Kepler grieved over his children's death.

2.3.6 *Galileo*

Kepler sent his book to many Astronomers in order to request a comment on the book. Two books sent to Italy were received by Galileo who had believed Copernican system and collected the proof concerning the movement of the Earth but concealing the collection.

Galileo sent a letter to Kepler on 4th August 1597 as follows:

"So far I have read only the introduction, but have learned from it in some measure your intentions and congratulate myself on the good future of having found such a man as a companion in the exploration of truth. For it is deplorable that there are so few who seek the truth. I will read your book in peace, for I am certain that I will find the most beautiful things in it. I would certainly dare to approach the public with my ways of thinking if there were more people of your mind. As this is not the case, I shall refrain from doing so." (Baumgardt, 1951)

Kepler replied in a letter on 13th October 1597 as follows: "Be of good cheer, Galileo, and appear in public. So great is the power of truth. If Italy seems less suitable for your publication and if you have to expect difficulties there, perhaps Germany will offer us more freedom" (Baumgardt, 1951). Galileo however kept his silence.

2.4 Tycho Brahe

2.4.1 *Tycho's comment on Kepler's book*

Kepler sent his book to Tycho Brahe wishing to know his comment on his book. Tycho did not evaluate the book as a high level book, although recognizing originality of Kepler's estimation. Tycho however told indirectly that Kepler could utilize the precise data of observation in Astronomy during his lifelong. At the time Kepler was aware that he should face Tycho.

2.4.2 *Religious suppression*

It occurred in Steiermark that Kepler was forced to go to Tycho. The Grand Duke Ferdinand II became to govern inner Austria including

Steiermark. His father the Grand Duke Karl agreed that there were Protestants in the duchy, but Ferdinand II neglected the concession of his father, and forced all people to be Catholic as his own religion, exercising the right according to reconciliation at Augsburg.

The Grand Duke ordered all the clergymen in the Protestant University, Church and school to retire from their works by 14 days later. Ten days later, the Bishop ordered all the Protestant clergymen and teachers to leave the city by one week later, and sentenced them to death if they did not obey his order. Kepler and his colleagues were banished, hurrying to collect necessities and leaving wives behind for farm, village. It was only Kepler who could be agreed to return to the city.

At the end of October of the year, since the entreaty requesting Kepler's return was agreed, Kepler returned to the city from the place where he hid. His friends and supporters had entreated Kepler's return because Kepler was not only a teacher but also the Mathematician for the district, and then he could return to the city as the Mathematician for the district.

In June 1599, the second children was born, but passed away by 35 days. Kepler rejected a Catholic burial, and hence he was imposed a fine on the rejection. He entreated a half fine, and he could not buried his child if never paying a fine. All the Bible translated by Ruter and heretic books were prohibited.

2.4.3 *Tycho's invitation*

After forced retire from the work of a teacher, Kepler escaped from reality to celestial estimation, and developed his idea concerning celestial harmony which would be published 20 years later. He however also searched a new work, but he failed to be appointed as a teacher in the University of Tubingen. Then, he knew that Tycho became the new Mathematician for Holy Roman Emperor Rudolf II, and lived in Praha. In December 1599, Tycho sent a letter for Kepler which invited Kepler to discuss the problem in Astronomy.

2.5 Astronomia Nova

2.5.1 *Praha*

On 11th January 1600, Kepler left Graz for Praha, and 10 days later arrived in Praha where there was the base of Holy Roman Emperor. Praha castle (Prazsky hrad) on the hill was the seat of power for Holy Roman Emperor. Praha was the capital of Bohemia. Tycho lived a castle in neighborhood of Praha because Emperor permitted Tycho to use the castle as he liked,

It was meaningful scientifically that Kepler met with Tycho on 4th February 1600. Two persons were very different, because Tycho was a aristocrat, had great self-confidence, being quarrelsome, and on the other hand Kepler was not an aristocrat, severe, thoughtful, never quarrelsome and never speaking with an air of importance. They got along together as key and lock nevertheless.

Astronomer Tycho collected the result of celestial observation during 35 years into the book consisting of 20 volumes. On the other hand, Mathematician Kepler published a thin book although based on estimation. Their skills made up each other. They both left their home. After quarrelling with the King of Denmark who was Tycho's supporter, he left home, living out of home. Kepler left Steiermark where there was an atmosphere of religious intolerance. Their meeting would yield the important change of Astronomy.

2.5.2 *Kepler's role*

Tycho was in the step where the precise theory of planets should be derived by analyzing the data of celestial observation during his lifelong, necessitating many assistants calculating the data. Tycho assigned Kepler to treat the theory of the Mars under Longomontanus (Christian Sorensen Longomontanus).

On 27th September 1600, a notice that all the citizens never converting to Catholic should be banished from the duchy, was issued in Graz.

Since Kepler rejected the conversion, he left Graz with his wife and daughter on two wagon carrying their property on 30th September 1600, arriving in Praha on 19th October.

When in the summer of 1601, Kepler returned to Steiermark, old Jobst Muller passed away. Then he returned to Graz in order to convert inherited asset of his mother in law to cash. When at the end of August Kepler returned to Praha, Tycho was planning that Kepler would be appointed as an assistant formally by the Emperor. In fact, there were few capable assistants excepting Kepler.

2.5.3 *Rudolfphine table*

Tycho who put confidence in Kepler, introduced Kepler to Emperor Rudolf II in the court. The Emperor was shy and eccentric. Tycho explained the planning to make a celestial movement table, and requested that it was permitted that the table would be named as "Rudolfphine table" for the Emperor. The celestial movement table was conventionally named for the supporter, and the supporter was assured that he could become an eternal invariant existence, satisfying the Emperor. It was only to pay wages for Kepler what Tycho wanted the Emperor to do.

2.5.4 *Tycho later*

On 13th October 1601, Tycho attended the banquet held at the house of Peter Wog Roznberg. Avoiding impoliteness he stayed at the banquet during time over the limit when the bladder could endure, yielding the fatal disease. He felt severe pain when he passed a little urine. The waste accumulated in his body and he suffered from heat. Considering his death, he said to Kepler "Please make efforts so as to find the proof that my life had never been fruitless."

On 24th October 1601, Tycho passed away. At the time, his celestial observation during 38 years ended. On 4th November, the funeral ceremony was performed. Two days later, Kepler became the new Mathematician for the Emperor, managed the equipment of Tycho, and was known that he should complete publishing Tycho's books which were not yet completed.

Fig. 2.7.   Rudolf II; Giuseppe Arcimboldo.

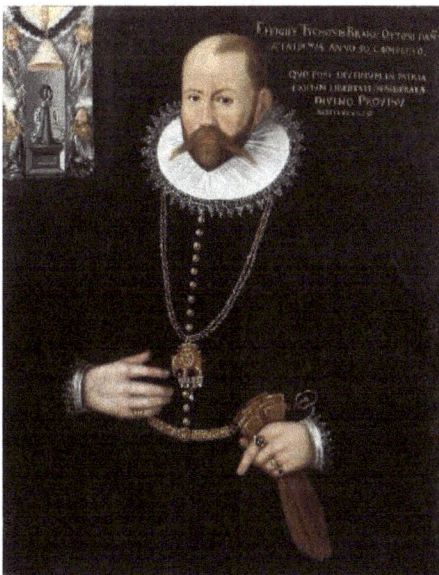

Fig. 2.8.   Tycho Brahe (1546–1601) (Portrait in 1596).

Explanation 2.2 Tycho Brahe

On 14th December 1546, Tycho Brahe was born at Schone in Denmark. His father Otte Brahe who was a Royal Privy Councilor, got married to Beate Bille who was a powerful figure at the Danish Court holding several royal land titles. Tycho was the eldest son among 12 brothers. When he was only 2 years old, he was taken away to be raised by his uncle Jorgen Brahe and Inger Oxe.

When he was 6–12 years old, he attended Latin school. On 19th April 1559, 12 years old Tycho started to study in the University of Copenhagen. He enrolled in the college of Law in accordance with the hope his uncle, and he however became interested in Astronomy.

On 11th November 1572, Tycho observed a strictly brilliant star in Cassiopeia, which is presently numbered as SN1572. Tycho found that the relative position between the new celestial body and the fixed stars in the background was not changed during several months, and then he considered that the new star was not a planet, but a fixed star on a celestial sphere far from every planets. Tycho created a term "nova" for the new fixed star.

From December of 1577 to January of 1578, the Great Comet appeared in northern sky. Tycho observing the Comet found that the distance between the Comet and the Earth was far away beyond the Moon, and the origin of the Comet could not exist in the celestial sphere of the Earth. He concluded that Aristoteles' theory insisting the invariant of celestial sphere far away beyond the Moon, was not correct. He also observed that the tail of the Comet extended always in the opposite direction of the Sun.

In 1588 Frederick II who was a supporter of Tycho, passed away. His son Christian IV became the successor, and then the support to Tycho ended. In 1599, Tycho got the support of Holy Roman Emperor Rudolf II, and moved to Praha as the Astronomer for the Court. In Praha he met Kepler and assigned Kepler to analyze the data of his observation during his lifelong. On 24th October 1601, he passed away.

2.5.5 *Research on the Mars*

Kepler believed that heliocentric system with the Sun as the center, existed in the matter created by the Creator and was the material symbol of the Creator. He considered that it was religiously important to prove the truth of the fact above mentioned. After Tycho passed away the research on the Mars by Kepler became physical. Kepler considered that all the Astronomy of planets would be clear by taking the Mars as a model.

A planet moved faster as it approached near the Sun. But it was a difficult problem how such a movement of planet above mentioned was described mathematically. Although in the case of circle orbit, the distance between planet and the Sun was constant, in the case of eccentric orbit where the Sun was separated from the center of orbit, the distance between planet and the Sun was changing during planet's movement on the orbit, also the velocity changing. Though modern mathematician could use differentiation and integration, at the days of Kepler these technique was not invented yet (there was need to wait until the days of Newton and Leipnitz).

2.5.6 *Law of constant area velocity*

Kepler used his own method instead of integral. When a planet moved on eccentric orbit from point A to point B, he calculated the distance between the planet and the Sun every one degree of angle, calculated the sum of such distances from point A to point B, and used the sum as a measure of time that it took to move from point A to point B. During this work, he remembered that Philosopher Archimedes in BC 2 century Greek calculated the area of a circle by using the sum of distances in the same way.

He considered that the area swept by the line between the Sun and a planet moving any distance on the orbit was well approximated by the

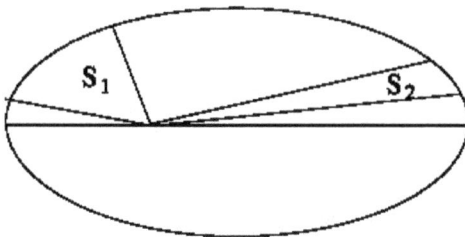

Fig. 2.9. Law of constant area velocity; S_1: area near the Sun, S_2: area far the Sun, $S_1 = S_2$.

measure which was the sum of distances between the Sun and a planet. As a result, Kepler arrived at the approximated principle "When the planet moves any distance on its path during the same interval, the area swept by the line between the Sun and a planet is the same."

Although the principle was derived by Kepler's original estimation never strict mathematical proof, the resulting content of the principle was correct. Hence this principle was called "the second law of Kepler" never called hypothesis.

2.5.7 *Elliptic orbit*

Kepler applied the law of constant area velocity to the circle orbit of planet with the Sun a little separated from the center of circle. He was aware that the time taken to move at any point on the orbit was too long, and the distance to the point on the orbit should be shorten. Hence he concluded that the orbit of planet should be an egg shape rather than a circle (Fig. 2.10).

It took 1 year of 1604 to solve the complicated problem specifying the type of egg shape appropriate, trying 20 types of shapes. As a result, He used an ellipse as egg shape. In this case, the error was opposite to the case of circle, because of egg shape too shortened. He considered that the correct orbit should be in the middle of them above mentioned orbits (Fig. 2.11).

The new ellipse had a feature that the Sun was placed at a focus of the ellipse. Since the distance between the Sun and a planet was calculated by the triangle method, a elliptic orbit could be derived simply. Thus Kepler

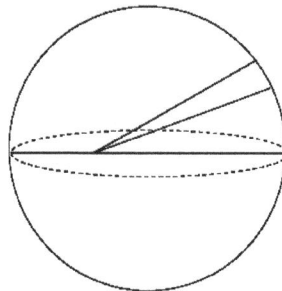

Fig. 2.10. Circular orbit; farther distance than real orbit.

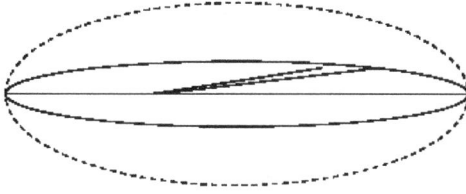

Fig. 2.11. Egg shape too suppressed.

arrived the first law that the orbit of a planet was an ellipse with the Sun placed at its focus. These two laws were expressed in "Astronomia Nova" published in 1609.

Later, these two laws were derived using strict physics and mathematics by Newton and were proved that Kepler's laws were correct. Hence these laws were called Kepler's laws never hypothesis.

2.6 Harmony

2.6.1 *Linz*

After Emperor Rudolf II passed away, there was nothing connecting Kepler to Praha. On the middle of April of 1612, Kepler left Praha for Linz. On 3rd September of previous year his wife Barbara who had suffered from infectious disease and passed away. In May 1612, Kepler arrived at Linz which was similar to Barbara's home Graz. The similarity was the reason why Kepler selected Linz.

Linz was the capital of the duchy upper Austria on the Northern-Eastern side of Steiermark, and almost citizens were Protestant. Here Kepler carried out the same work as in Graz, and he continued to make "Rudolfphine table" and made a map of upper Austria. He also was a teacher of school in the district and a Mathematician for the district.

2.6.2 *Celestial harmony*

In "Astronomia Nova," the Mars was researched. In 1614 Kepler researched other planets than the Mars. In May 1616, he became to start mass-producing ephemeris which included the table indicating the

Table 2.1.   Kepler's third law.

	Mer.	Ven.	Ear.	Mar.	Jup.	Sat.
Period	0.24	0.616	1.00	1.88	11.87	29.477
Distance	0.388	0.724	1.00	1.524	5.20	9.51
$\dfrac{\text{Period}^2}{\text{Distance}^3}$	0.99	1.00	1.00	1.00	1.00	1.01

Note: The unit of period: year, the unit of distance: the average distance of the Earth from the Sun (Voelkel, 1999).

position of each planet every day of the year, and indicating also the position and the time of estimated phenomenon such as solar-eclipse, lunar eclipse and appearance of celestial bodies. The ephemeris was a important necessary book for mate and astrologer.

At the end of December 1615 when Kepler was working for ephemeris, he heard that his 68 years old mother took criticism that she might be a witch. The relatives sent her to him in Linz for living with her, in order to prevent his mother from making something more wrong.

It was the aim which he had wished to achieve at the change from Church to Science in his life, what Kepler came back to research in 1618. It was that he praised the glory of the Creator by expressing the mathematical regularity existing in Nature. It was the problem next to two Kepler's laws above mentioned how all the planets were in harmony with each other. Kepler made efforts with insistence until he got the reasonable solution.

Researching the relation between the period of planet orbit and the average distance of the planet from the Sun, Kepler found that the ratio of the square of the period to the cube of the average distance was the same (Table 2.1), called "The third law of Kepler", which was expressed in *Harmonice Mundi* published in 1619.

Explanation 2.3 Kepler's laws

(a) Property of ellipse
To understand Kepler's laws, properties of ellipse are explained in the following. Semimajor axis a and semiminor axis b are defined as Fig. 2.12.

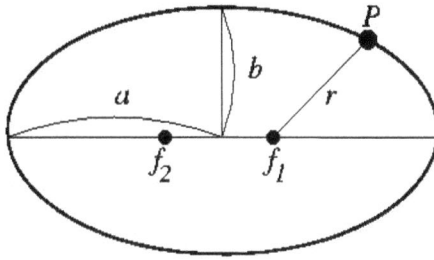

Fig. 2.12. Semimajor axis a and semiminor axis b: f_1 and f_2 are foci.

Two foci f_1 and f_2 are defined as Fig. 2.12. Distance r between f_1 and point P placed by a planet on ellipse, is defined as Fig. 2.12. Distance r at perihelion is defined as r_2. Distance r at aphelion is defined as r_1. Semimajor axis a is expressed by $(r_1 + r_2)/2$ and semiminor axis b is expressed by $(r_1 \times r_2)^{1/2}$ where $(\ )^{1/2}$ expresses square root.

Semimajor axis was used in Kepler's third law. Perihelion appeared in "the perihelion motion of the Mercury" which was used to verify the correctness of the general relativistic theory (Explanation 6.8).

(b) Kepler's first law
Kepler analyzed enormous amount of precise data of orbital motion of planet by Brahe. Consequently concerning the motion of the Mars he found that the planet moved on an elliptic orbit around the Sun, and one of the foci was the place of the Sun. This fact was found to be correct for other all planets. This is the Kepler's first law that is, "A planet moves on elliptic orbit which has a focus placed by the Sun."

(c) Kepler's second law
Furthermore, Kepler discovered Kepler's second law that is, "area which is swept by a line between a planet and the Sun, is same when a planet moves anywhere on an elliptic path during same interval." In 1609, this was described in *New Astronomy* published in Heidelberg.

Afterward, by Newton it was proved that if gravity on a planet was proportional to inverse square of distance, then Kepler's second law was

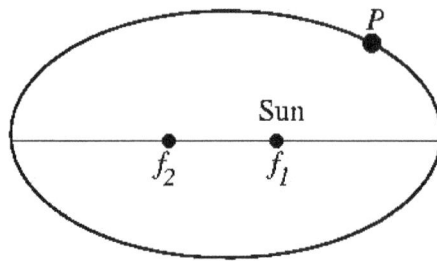

Fig. 2.13.   Kepler's first law.

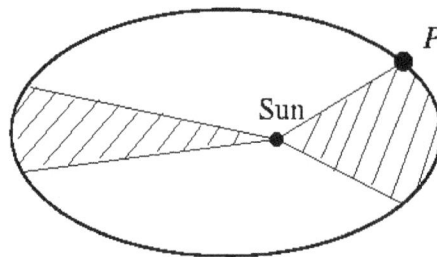

Fig. 2.14.   Kepler's second law.

derived. That is, Kepler's second law was helpful to prove that if gravity obeyed the inverse square law of distance then a planet moved on an elliptic orbit (Explanation 3.6).

(d) Kepler's third law

In 1619, concerning any two planets, Kepler described Kepler's third law that the square of ratio of orbit's periodic times, was equal to the cube of ratio of semimajor axis, in *Harmony of the World (Harmonices mundi libri)* published in Linz. Kepler's third law implied the square of orbit's periodic time being proportional to the cube of semimajor axis.

By Newton it was proved that if Kepler's third law held true then gravity on a planet was proportional to the inverse square of distance from the Sun. Kepler's third law was helpful to prove that if a planet moved in an elliptic orbit then gravity obeyed the inverse square law of distance

(Explanation 3.7). In analyzing data of the Mars for discovering three laws, Kepler used about one thousand papers for calculation.

2.7 Kepler Later

2.7.1 *The beginning of 30 years war*

On 3rd October 1621 in the court's decision of witch trial the Duke of Wurtenberg ordered to liberate Katharine Kepler. The court ruled that she was not guilty after Kepler's efforts for his mother. This trial claimed her energy and she passed away 6 months later on 13th April 1622.

On 23rd May 1618, dissatisfied Protestant delegates made a raid on Privy Council's conference room in Praha castle, and threw out regents from window, called "Defenestration of Praha," yielding the beginning of 30 years war. However in November 1621, Protestant revolution which had occurred in the area along the Danube in Bohemia and Austria, was

Fig. 2.15. Defenestration of Praha.

suppressed, and 27 Protestant masterminds of resistance in Praha were executed.

It was unexpected that Kepler in Linz was appointed as the Mathematician for Empire by Catholic Ferdinand II on 30th December 1621. Even when all the citizens at Linz were ordered to convert to Catholicism, not only Kepler but also printer Plank whom Kepler utilized, and Plank's assistants were permitted to stay in Linz. In 1622, Kepler became to the end of working for "Rudolfphine table" to which Kepler devoted himself during long time.

On 24th June 1626, a big army of farmers surrounded Linz, and set fire to the edge of the city. Plank's printer burned though without damage of "Rudolfphine table." Since there was not a reason that they stayed in Linz. Kepler sent a letter requesting to permit that they left Linz. On 20th November he got on a small boat with his family, and sailed upstream the Danube searching the place where he could print "Rudolfphine table."

2.7.2 *Regensburg*

The boat where Kepler's family were getting on, sailed upstream the Danube to Regensburg, but the river was frozen beyond there. Hence Kepler decided for his family to stay in Regensburg as a rcfuge, and he left alone for Urm loading types on the wagon. On 10th December 1626, Kepler stayed opposite side of Jonas Saul's printing office. "Rudolfphine table" was printed by September 1627. On 15th September, Kepler left bringing his books for Frankfurt where there would be an autumn market of book. At the end of November, he returned to Regensburg.

On 8th October 1630, Kepler left for Leipzig where an autumn market would be held. After visiting the market Kepler had the coachman go ahead to Regensburg, and several days later he left for Regensburg. On 2nd November, he returned to Regensburg feeling cold and very tired. He got sick because he travelled in a cold air of autumn. He had a fever and fell unconscious. On 15th November 1630 he passed away. Two days later, he was buried at a graveyard of St. Petro Church in outside of castle wall of Regensburg.

Several years later, Regensburg was offended by the army of Sweden, and the army of Empire defended. During the battle Kepler's tomb was erased.

The only record concerning Kepler's tomb was the sketch which was drawn by his friend. In the sketch, there was the explanation on Kepler's career that he was the Mathematician for three Emperors, and also he was the most magnificent Astronomer.

The discoveries by Galileo and Kepler contributed to foundation of Newtonian mechanics by Isaac Newton who inheriting Galileo and Kepler elucidated the movement of celestial bodies of the universe.

Chapter 3

Isaac Newton

Isaac Newton (1643–1727). Sketch by author.

By motion's laws discovered by Isaac Newton, motion of celestial bodies was elucidated. Newtonian mechanics founded by him constituted the greatest two theories of classical physics with electromagnetic theory which was theorized by Maxwell on the basis of experimental researches on electromagnetic phenomena by Faraday.

3.1 Upbringing

3.1.1 *Birth of Newton*

Newton was born on 4th January 1643 (on 25th December 1642 in Julian calendar) at manor house in Lincolnshire Woolsthorpe in one year after Galileo who was the greatest thinker up to Newton's time, passed away. His father Isaac Newton senior was a farmer and inherited manor and manor house, and was lord with seigniorial authority over a handful of tenant farmers. The lord like this was called as yeoman then. He passed away of disease three months before the birth of Newton.

Newton's mother (Hannah) remarried with the rector (Barnabas Smith) at North Witham neibouring village at his age of 3. Stepfather Smith required that Newton would not be brought to new home. Therefore Newton was brought up by his grand mother (Margery Ayscough).

Social background then was the following. Charles I was beheaded in 1649, and Puritan Revolution started. In 1653, Cromwell (Oliver Cromwell) became Lord Protector. Until 1658, civil war between Puritans and Royalists was continued. Even in countryside, soldiers of Puritan pursued Royalists, and the political situation was unstable.

3.1.2 *King's school admission*

Because stepfather passed away in 1653, Newton lived in large family with grandmother, mother, one brother who was a son of his mother and stepfather and two sisters who were daughters of his mother and stepfather. One year after living in large family, Newton entered King's school (founded in 1528) at Grantham. This school was evaluated as a school preparing to take the entrance examination for Oxford University and

Fig. 3.1. The manor house where Newton was born, in Woolsthorpe (in June 2016 taken by author).

Cambridge University. Grantham was a market town of a few hundred families and a key point in Lincolnshire which was important distribution center of agricultural products.

Grantham was seven miles away from Woolsthorpe — a distance much too great to walk to school each day. Newton became a lodger with the family of the Clarks who was the apothecary. Second wife of Mr. Clark was a friend of Newton's mother. Catherine (Catherine Storer) who was children from his wife's previous marriage, was 2 years younger than Newton and was merry. She eased tension of modest Newton who came from countryside. Afterward they would become engaged couple.

3.1.3 *A fight*

Newton was not interested in studying at school, and was in the lowest grades. Then his only interest was that he received his earliest knowledge of primitive chemistry from Clark. One day, he was kicked hard in the stomach by a classmate in superior grade. After school at church yard Newton challenged the much larger boy to a fight. During their fight, the schoolmaster's son who disliked Newton's antagonist, came to them and said "Hey boys! Newton is beginning a fight." Hence Newton was in the public eye. Though Newton was not so lusty as his antagonist, he fought with so much more spirit and resolution, and beat him until the boy declared to no more fight.

Upon that time, the schoolmaster's son bad Newton use antagonist like a coward & rub his nose against the wall & accordingly Newton pulled him along by the ears and thrust his face against the side of the church. Still not content, before leaving the bully to nurse his wounds, Newton declared he would not rest until he had overtaken his combatant's academic position (White, 1998:22). Since this episode, Newton within a short time, rose to first place in the school. He was so interested in learning that the schoolmaster was surprised.

3.1.4 *Two years leave of absence*

His mother got her wealthy living because of increasing income from manor. She decided that she entrusted management of manor to the eldest son. Therefore, she applied for two years leave of absence from 1658, and made him work farm, although he rose to higher grade. However, he was lost in thought frequently, 2 or 3 hours after beginning of work. He was unfit for working farm.

Newton's uncle Ayscough (William Ayscough) was the rector of Anglican church after graduating from Cambridge University. The uncle persuaded his mother that Isaac would be back in Grantham in 1660.

The restoration occurred with Charles II taking the throne. Political situation became peaceful.

3.2 Cambridge University Admission

3.2.1 *Sizars among four categories*

On 5th June 1661, Newton enrolled at Trinity College of Cambridge University. Among 40 classmates, almost all of the students were youths in superior social position. They had prepared in public school, the entrance at Cambridge University, and were wealthy. Cambridge recognized students in four categories: fellow-commoners, pensioners, sizars, and subsizars. Fellow-commoners were privileged students (White, 1998:96), were dressed in sophisticated gowns and dined at high table (Gleick, 2003:20). Pensioners paid tuition fees and boarding fees and aimed for the rector. Sizars were exempted from tuition fees and boarding fees (subsizars paid tuition fees), and paid their way by emptying the bedpans, cleaning the rooms and running errands of the more privileged students (White, 1998:46). Sizars waited on student who received menial service, at meals and ate their leftovers. Newton entered first as subsizars and presently became sizars.

Though Newton's mother was in a wealth state, she gave little money as school expenses because she hoped for him to manage manor, and was not interested in scholarship. Therefore, Newton should enter at university as sizars or subsizars. Although Newton should provide menial service to privileged student, he enrolled at Cambridge University as sizars because he yearned for scholarship.

Fig. 3.2. Trinity College of Cambridge University; Sketch by author.

Position as sizars might be treated with contempt by those in superior social position, or might be ignored. This made him be an introvert. In Trinity Hall two students shared the room. His room mate was in superior social position and had many friends. When the friends visited the room, the room was noisy and he could not concentrate his attention on his study. Then, he was quietly thinking viewing nocturnal sky frequently in courtyard. Another student who had the same distress, happened to be in the courtyard, and proposed to negotiate with university for him to share the room with Newton. Because the proposal was realized, Newton and the student could study quietly.

3.2.2   *Curriculum*

When Newton entered at Cambridge University, the curriculum of university took over the middle Ages. The contents of education were theology, classics, law and medicine, especially theology and classics were treated seriously. Natural science including mathematics did not exist in curriculum. As mentioned below, Newton self-studied mathematics. The traditional backbone of university was the old notions of Aristotle, and logic, ethics and rhetoric were the basis of philosophy. However, then, at university in the Continent, radical ideas of Galileo, Descartes (Rene Descartes) and Kepler (Johannes Kepler) were paid attention. At the library of College, Newton studied on ideas of Descartes, Galileo and Kepler. Newton immediately studied Descartes philosophy. Inserting the name of Aristotle into the word of Aristotle, he noted down "I am a friend of Plato, I am a friend of Aristotle, but truth is my greater friend" (Gleik, 2003:26). This note makes us imagine Newton's future.

3.3 Academic Development in the Continent Focusing on Astronomy

Academic development in the Continent which constituted the background of Newton's discovery, is described focusing the attention on astronomy in the following.

3.3.1 *Kepler*

First in1543, Copernicus (Nicolaus Copernicus) published *On the Revolutions of the Heavenly Spheres* (*De revolutionibus orbium coelestium*). In this book, he described the heliocentlic theory where the Sun was set at the center of universe and the Earth revolved around the Sun. As supporters for him, Galileo and Kepler who analyzed the data of astronomical surveys by Brahe (Tycho Brahe), appeared.

Brahe contributed beyond measure to astronomy and brought revolution to instrument for astronomical surveys. Simultaneously he introduced a revolutionary method in astronomical surveys. For example, on observing a planet, though till then planets were observed only at special time, he continuously observed an orbit of planet, and got the vast body of remarkably precise data of astronomical surveys.

When Newton thought what controlled orbital motions of celestial bodies, the knowledge concerning how planets moved, was the important basis on which the fundamental principle was researched. The laws of orbital motion of planets were discovered by Kepler who analyzed the vast body of data of astronomical surveys by Brahe.

3.3.2 *Galileo*

Newton was greatly influenced by Galileo's work when he understood physical phenomena. Newton generalized the basic thinking of the realm of motion by Galileo, and unified theories of Galileo and Kepler, and founded Newtonian mechanics as theory of motion.

3.4 Barrow, the Lucasian Professor of Mathematics

3.4.1 *Descartes*

The person whom Newton was interested in, was Descartes. Descartes applied algebra to geometry for the first time. In 1637, Descartes published *Discourse on the method* (*Discours de la method pour bien conduire sa raison et chercher la verite dans les sciences. Plus la Dioptrique, les*

Fig. 3.3.   Isaac Barrow (1630–1677).

Fig. 3.4.   Statue of Isaac Barrow (taken by author in June 2016 at Trinity Chapel).

Meteores et la Geometrie, qui sont des essais de cette method) at Leiden in the Netherlands. The book consists of three papers, and the preface is called "Discourse on the method" and accompanying three papers are *La Dioptrique*, *Les Meteores* and *La Geometrie*. Newton read separate volume translated into Latin of *La Geometrie* with enthusiasm.

In 1663, Newton bought the book of Astronomy at market in square. However, he could not understand the mathematics in the book and he became aware of lack of knowledge of geometry. He decided to read "Elements" by Euclid. He studied the geometry of Descartes, and new algebra and analytical geometry. Studying mathematics, he devised his own proof method different from the author's method.

3.4.2 *Barrow*

In 1663, when Newton began to study mathematics, Barrow (Isaac Barrow) who was a mathematician and a theologian, was inaugurated as the Lucasian Professor of Mathematics. He delivered lectures of natural philosophy (then, science was called thus) and optics. Newton attended the lecture of Barrow. Barrow was a good teacher who found genius of Newton and trained him.

Though Newton had not extra money because he was a sizars, he brought humble gifts to Catherine at Granthem where he had lodged, and his brother and sisters who were the son and daughters of his mother and stepfather. He was a tender youth who took care of his friend and his brother and sisters.

Barrow evaluated the creativity of Newton who was 12 years younger than him and an innocent youth without a lust for fame. He thought that he should look after Newton until he would splendidly grow up because he would surely become a great person someday. Receiving the respect of Professor, Newton studied hard.

3.5 The Highest Creativity in Birthplace Under the Situation Due to Plague

3.5.1 *Plague*

In 1665, Newton got a Bachelor of Arts from Cambridge University. In the summer, the plague spread quickly in London. University was closed and he went to his home in Woolsthorpe. The 2 years in his birthplace was the period when he showed the highest creativity in his life. He treated the three important problems concerning physics-astronomy, optics and mathematics. Consequently, he accomplished revolutionary discoveries and inventions in those realms.

3.5.2 *Falling apple*

First, Newton had been successful in research on physics and astronomy. 24 years before the birth of Newton, Kepler published laws of orbital motion of planets in "New Astronomy" and "Harmony of the World" (Paragraphs 2.5 and 2.6) as an answer to the question how planets moved. However, it was not elucidated why planets should move following such laws.

One day, Newton was lost in thought in orchard. The topic that then an apple fell near him, is due to biographer Stukeley (William Stukeley) who in the spring of 1726, visited Newton before Newton passed away. "Free fall" expressed by falling of body on the ground such as an apple, was researched in detail by Galileo (Explanation 1.3).

Then, Newton thought "Is the Moon influenced by the same gravity as the gravity which influences an object like an apple on the ground?" He had a question "If the gravity of the Earth influences the Moon, why does the Moon not fall on the ground as an object like an apple falling on the ground?"

To this question Newton thought the following: if the gravity of the Earth did not influence the Moon, by inertia's law the Moon moved linearly in direction of velocity and flew away a long way off in universe. The fact that without flying away the Moon revolved around the Earth, showed that the Moon fell to the Earth from point a to point b as Fig. 3.5. He calculated the distance of the falling of the Moon during one second and noted down it. During the following 20 years, it was not disclosed (White, 1998:92).

Furthermore, Newton thought what law should be satisfied by the gravity between the Sun and planet, in order for orbital motion of planet to satisfy Kepler's third law, where the square of periodic time of planet's

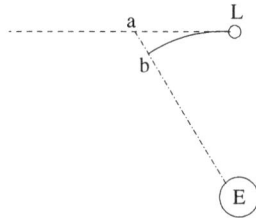

Fig. 3.5. Drop of the Moon (L) due to gravity of the Earth (E).

orbital motion was proportional to the cube of semimajor axis of the ellipse with focus placed by the Sun. Consequently, he was successful in discovering the inverse square law of distance that is, "Intensity of gravity was proportional to the inverse square of distance" (Explanation 3.7).

All truth did not occur to him as divine message in an instant. Newton said himself "I kept the subject of gravitation constantly before me till the first dawnings opened slowly little by little into the full and clear light."

3.5.3 *Rainbow by prism*

In research on optics, Newton used a triangle prism on experiment. When sunbeam was passed through prism, light was refracted and beautiful rainbow color of light was observed. From the observation he concluded "white light did not consist of the single light, but consisted of many different colors of light." He discovered "Different color of light was refracted differently" (Explanation 3.1).

Explanation 3.1 Refraction of light through prism

The incident light with incoming angle ϕ is refracted at B with refraction angle Ψ, and go through to C where the light with angle Ψ' is refracted with refraction angle ϕ', and go away (Fig. 3.6). Let ε be the bending angle which is the bending angle between incident line and outgoing line, and the refractive index n is given by

$$n = \sin\phi/\sin\Psi, \ n = \sin\phi'/\sin\Psi'$$

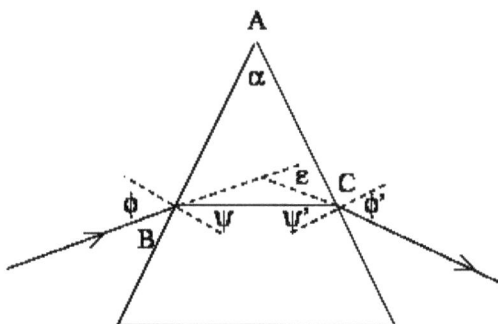

Fig. 3.6.   Refraction by prism.

From geometric relation, we have

$$\Psi + \Psi' = \alpha, \; \phi + \phi' = \varepsilon + \alpha$$

where α is the angle at point A of prism. When $\phi = \phi'$ and $\Psi = \Psi'$, ε has the minimal value ε_{min}, then we have

$$\varepsilon_{min} = 2\phi - \alpha, \; 2\Psi = \alpha$$
$$n = \sin[(\varepsilon_{min} + \alpha)/2]/\sin[\alpha/2]$$

Hence we can obtain n by observing ε_{min}.

3.5.4 *Differentiation and integration*

In research on mathematics, Newton invented the methods calculating "differentiation and integration," under the influence of "Geometry" by Descartes, gradient and curve studied from Professor Barrow. Independently, Leibniz (Gottfried Wilhelm von Leibniz) invented the methods calculating differentiation and integration. However, invention by Newton was several years earlier than Leibniz's invention. Newton called his method mentioned above as "method of fluxions." He thought that the integral calculus was the inverse of the differential calculus. Regarding differentiation as elementary operation, he created the

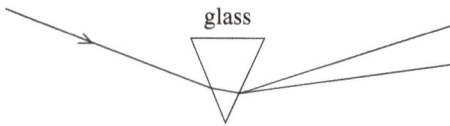

Fig. 3.7.   Refraction of light with prism.

Fig. 3.8.   Gottfried Wilhelm von Leibniz (1646–1716) (portrait in 1695).

analytical method unifying different techniques calculating such as area, tangential line, arclength of curve, maximum and minimum of function.

3.5.5 *Reopen of University*

Because plague declined, in 1667, Cambridge University reopened. In March of this year, in the College chapel, Newton was successful in oral questioning and a written test (White, 1998:95). In October, he became a Minor Fellow of Trinity College. A Minor Fellow was provided a stipend and an allowance. It was most important that he could continue to research the previous subjects. He was given a room free of charge.

Though he was engaged to Catherine of the family of the Clarks with whom he had lodged, the engagement was dissolved under agreement because she estimated that Newton would become a Minor Fellow. A new

Minor Fellow was forbidden to marry for 7 years. It was the reason why their engagement was dissolved. Newton did not forget her in life. Afterward, she lost her first husband. After that, she remarried, but became again widow. For her in such situation he did not spare economical support.

Explanation 3.2 Principle of telescope

In Fig. 3.9, when sufficient far object B is observed, incident light is parallel, and the image of B results at the place of focal length f_o after lens L_1. When B′ is placed at focal length f_e before lens L_2, lights in projection space far outgoing lights from B′ are parallel, and the image for B′ is observed as a virtual image. Let the angle of incident light be θ, and let the angle viewing the virtual image be θ', then the magnification of telescope is given as

$$\text{Magnification} = \tan\theta' \,/\, \tan\theta = -f_o/f_e$$

where negativity of magnification means inverted image.

At the part near edge of ⌘ lens with remarkable curvature, light passing through is sharply bended. Light passing through near center with the least curvature of lens is slightly bended, Consequently unclear image results. This problem is called "spherical aberration." In Kepler's telescope, there was a problem of the spherical aberration.

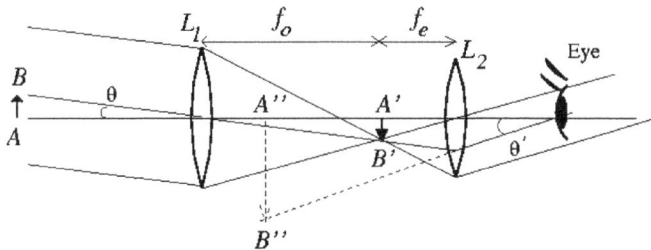

Fig. 3.9. Image of telescope.

3.5.6 *Going home to report a Minor Fellow*

In order to report that Newton became a Minor Fellow, he went home in Woolsthorpe for a time. Then, he talked about telescope to his brother and sisters. Since telescope was invented in the Netherlands, Galileo used ⟆ lens as object lens and used ⟅⟆ lens as ocular lens, and improved the magnification by polishing lens to adjust focal length. Kepler spreaded the range of view using two ⟆ lenses. However, due to spherical aberration where the bending of the light was different at center or edge of lens, a distorted image resulted. To prevent from the problem, Newton devised the different type of telescope. That is, as Fig. 3.10, he set ⟅⟆ mirror at the bottom of cylinder, and in order to observe at the side of cylinder, set a mirror using ⟆ lens as ocular lens. The magnification of the reflecting telescope by Newton was 40. The huge reflecting telescope set at astronomical observatory in present world, goes back to Newton's reflecting telescope.

In 1668, Newton came back to Cambridge from his home, and got Master of Arts. In March, he became Major Fellow and a stipend and an allowance were increased.

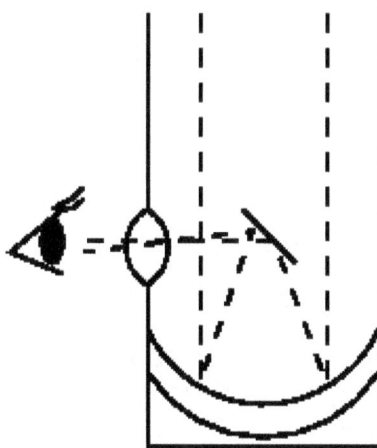

Fig. 3.10. Newton's reflecting telescope.

3.6 Successor of Barrow

3.6.1 *The Lucasian Professor of Mathematics*

In 1669, his former teacher Barrow sent Newton's manuscripts to Collins (John Collins) who was the chief librarian of the Royal Society, in order to inform Newton's research works on mathematics. Collins immediately informed contents of the manuscripts to noted mathematicians. Also, he sent the manuscripts to the President of the Royal Society Brouncker (William Brouncker) in agreement with Newton. However, afterward, Newton who had not ambition of standing out in academic society, and had not a lust for fame, asked return of manuscripts. Consequently, the detail of Newton's research works on mathematics was left unclear. However, most mathematicians understood outline of Newton's research works through Collins.

In 1669, Barrow retired from the Lucasian Professor of Mathematics in order to devote himself to theology, and nominated Newton as the successor.

In October 1669, Newton was inaugurated as the Lucasian Professor of Mathematics.

As the Lucasian Professor of Mathematics, the first lecture was delivered in January 1670 (White, 1998:163). Newton delivered lecture profoundly thinking. The lecture was so high level to understand. Not a single student showed up for Newton's lecture. For the lecture which was especially difficult to understand, many students stayed away from the lecture. He did not lower the lecture without playing up to students. When he went through the courtyard and reached lecture room and did not find any student, he waited during 15 minutes till some student attended, in case of no student attending he went back to his room. In his room, he continued his research. The habitual absence of students continued throughout almost every lecture for the next seventeen years (White, 1998:164).

On the other hand, as the Lucasian Professor of Mathematics he first researched on optics. Since Aristotle, all scientists thought that solar light consisted of the single element. However, chromatic aberration at lens of telescope led Newton's conclusion different from previous common knowledge. As mentioned above, Newton was aware of that white light passing through prism was observed as split lights with plenty of clours from red

Fig. 3.11. Newton's reflecting telescope (replica owned by the Royal Society) (@ Andrew Dunn (lisensed under CC BY SA 20) (https://creativecommons.org/lisenses/by-sa/2.0)).

to violet. That is, Newton discovered that white light was composed of colours, and a light with different color was refracted in different angle.

3.6.2 *Presenting his reflecting telescope*

In December 1671, Newton presented his own reflecting telescope to the Royal Society. His reflecting telescope with high magnification was paid attention, and his name was known in London. In January 1672, after the presentation he was elected as a Fellow of the Royal Society.

When Newton's reflecting telescope was demonstrated in the Royal Society, Hooke (Robert Hooke) who was Curator of Experiments in the Royal Society, unsatisfactorily said "Though I tried to fabricate reflecting telescope, but could not complete it because of plague." It was notified by Collins (White, 1998:178). Afterward, Hooke criticized Newton's papers (as mentioned below).

3.6.3 *The first published paper on optics*

In 1672, Newton submitted the research result on the composition of white light to Oldenburg (Henry Oldenburg) secretary of the Royal society.

It was carried as his first paper on "Theory of Light and Colors" in the Philosophical Transaction of the Royal Society. This paper was accepted favorably.

In this paper, Newton explained the experimental fact "White light was composed of plenty of elements being refracted differently, and was split into different colors of light. Prism did not add color to light." He thought that light was stream of particle. He thought that sound wave showed diffraction but light went straight, and consequently light was not wave. However, Huygens (Christiaan Huygens) who insisted "light being wave," and Hooke who was in position of refereeing paper, criticized Newton's idea of "light being corpuscular." Hooke insisted that color was added to light by prism. Newton objected in the following: "paper is the scientific truth based on the experimental facts. It is fault that they ignore the results of my experimental fact because my idea is contrary to their idea," (Sootin, 1955). After all, the idea of "light being corpuscular" by Newton who got good reputation, dominated, and its superiority continued till the idea "light being wave" revived in 19th century.

Newton regretted that he published paper. Because he thought that he wasted important time disputing with critic about his paper, he thought the following: "I do not desire neither reputation nor fame. Such desire lets a person waste his important time in life. If my presence is not known, then I can use sufficient time for thought" (Sootin, 1955).

In March 1673, he sent a letter which informed to resign Fellow of the Royal Society, to Oldenburg. The patient persuasion by Oldenburg prevented from resigning by Newton.

The idea of "light being wave" and "light being corpuscular" are both correct because at present "the particle-wave duality of light" (Appendix 3.1) is verified.

Appendix 3.1 The corpuscular-wave duality of light

In the 19th century, Maxwell derived 4 electromagnetic equations, and indicated that electric field and magnetic field satisfied the wave equation. He predicted theoretically the presence of electromagnetic wave, and predicted that light was identical to electromagnetic wave. This prediction was afterward verified experimentally by Hertz (Heirich Rudolph Hertz).

In experiments from 1886 to 1889, Hertz made electromagnetic wave with high frequency electric vibration, and verified that in electromagnetic wave, refraction, reflection, and polarization occurred as light experimentally. He concluded that light was identical to electromagnetic wave. On the other hand, the idea "light being corpuscular" was proved in the following. In 1900, Planck (Max Karl Ernst Ludwig Planck) who played the role of leading the beginning of quantum mechanics, proposed "energy of light is expressed as the unit called quantum multiplied by integer, and energy quantum is proportional to frequency of light." Light with energy of quantum was called photon. In 1905, Einstein (Albert Einstein) proved light being corpuscular, by the theory on photoelectric effect (Explanation 3.3). Thus, the idea "light being corpuscular" was proved.

Explanation 3.3 The photoelectric effect

As figure, when light is radiated on metal (M), photoelectron (e) is emitted. This phenomenon is called "photoelectric effect." In Fig. 3.12, ν is frequency of light. Planck proposed that $h\nu$ is energy quantum (Appendix 3.1). h is Planck constant. Einstein called light with energy quantum as "photon" (Appendix 3.1).

Properties of photoelectric effect are expressed in the following.

① If the light with longer wave length than limit wave length λ_0 is radiated, then photoelectric effect does not occur.
② Energy of electron is dependent on frequency of light and is irrespective of amplitude of light.
③ Number of electrons is proportional to intensity of light radiated.

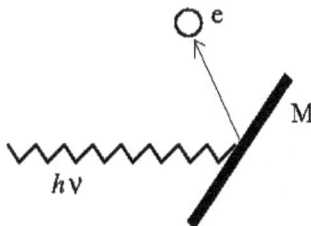

Fig. 3.12.   The photoelectric effect.

Explanation on photoelectric effect by idea of "light being wave" is difficult as follows.

(1) If light is wavelike, photoelectric effect may be considered to occur by radiating light with strong intensity. But the property ① is contrary to the above mention.
(2) If light is wavelike, then electron with more energy may be considered to be emitted for irradiated light with more intensity. But the property ② is contrary to the above mention.

In the beginning of 20th century, Einstein elucidated theoretically the photoelectric effect (Appendix 3.1 and Pragraph 6.4).

3.6.4 *The sorrow due to precious persons*

In the latter 1670s, precious persons for Newton passed away. In 1677, Barrow and Oldenburg passed away. Afterward, hearing critical illness of his mother, Newton went home to nurse his mother. Consoling his mother suffering from pain he sat up whole nights with her (White, 1998:193). In spite of his devoted nurse, she passed away. The sorrow and loneliness due to losing his mother were continued for a long time.

Newton was unsatisfied with her death being recorded officially in the name of Barnabas Smith. However, Newton was satisfied with seeing his mother buried beside his natural father, rather than his despised stepfather. After he ended the management work for inherited estate, he came back to Cambridge and kept away from person, devoting himself to theology.

3.7 Principia

3.7.1 *Motion's three laws*

In 1687, Newton published *Mathematical Principle of Natural Philosophy* (*Philosophiae Naturalis Principia Mathematica*). This book is known as *Principia*. In this section, we describe the process of publication of *Principia*.

   Principia was composed of Introduction, Books I, II and III. Before 1666 Newton had early idea of "Motion's three laws" (Explanation 3.4) described in the Introduction. The first law: "inertia's law," the second law: "motion's law," and the third law: "action–reaction law" were the three laws.

   "Inertia's law" was useful as follows. As mentioned in Paragraph 3.5, if the Moon did not receive force from the Earth, then obeying to "inertia's law" the Moon moved with constant velocity in linear direction, and flew far away to universe. But, without flying away, the Moon revolved around the Earth. The reason why the Moon revolved around the Earth, was that the Moon received some force from the Earth. Thus, "inertia's law" played important role in his discovering universal gravitation.

   "Motion's law" was useful as follows. Newton already discovered the centrifugal force (Explanation 3.5) influencing a body revolving circularly. The fact that the Moon revolves around the Earth, indicates that the Moon has acceleration because of no linear motion with constant velocity, and receives some force due to "motion's law." This force is the centrifugal force, and equilibrates with the gravity of the Earth. Consequently, the Moon does not fall on the ground. The above mention was based on the motion's law.

Explanation 3.4 Motion's three laws

Inertia's law: "When a body does not receive force or two forces equilibrate, a body is at rest or move at constant velocity."

   The first law "inertia's law" is a law explaining the property that a body does not change the velocity when a body does not receive force. Under the law, when a body does not receive force, a body at rest keeps at rest, and a body moving at a velocity moves at the same constant velocity. For example, when a train is braked, passengers fall in moving direction. This is because a train receiving force is decrease the velocity, but passenger keep the previous velocity. That is, the above mention is the result of inertia's law.

   Under "inertia's law" a body moving differently from motion with the constant velocity and in linear direction, receives force. Consequently, the

Moon revolving around the Earth which moves differently from motion with constant velocity and linear direction, receives the gravity of the Earth. Thus "inertia's law" played the important role of discovering the universal gravitation by Newton (Paragraph 3.5).

Motion's law: "When a force influences a body, the product of resulting acceleration by mass of body is equal to the force."

   The second law "motion's law" describes motion equation, that is, "force is given by product of mass by acceleration." Acceleration is the differentiation of velocity, and expresses the change of velocity per unit time. Thus Newton expressed motion's law with his own invention of differentiation. By the law, it was expressed that adding force to a body produces acceleration and the acceleration is proportional to force, and to the inverse mass. For example, the motion's law is the reason why the force necessary for moving at a velocity from at rest is larger in case of heavy load than in case of vacancy. In order to obtain the same acceleration, larger force is necessary in case of heavy mass.

Action-reaction law: "When body 1 influences force F to body 2, body 2 also influences the force $-F$ with the same magnitude and reverse direction to body 1."

   The third law "action–reaction law" expresses the law that if some force influences, then the force with reverse direction influences. For example, when a body is placed on a floor, the gravity influencing a body equilibrates with the resistive force, consequently a body can exist stably on the floor. If the floor is not substantial, the resistive force does not influence a body because of floor falling, and the body on the floor loses balancing.

3.7.2 *Centrifugal force*

Centrifugal force is proportional to the product of the square of the angular velocity by the radius of circular motion being the distance between the Earth and the Moon, (Explanation 3.5) and the angular velocity is inversely proportional to periodic time. Consequently the centrifugal force is proportional to the inverse square of distance between the Earth and the

Moon due to Kepler's third law. The gravitation equilibrating with centrifugal force is inverse in direction and has the same magnitude as the centrifugal force, and so proportional to the inverse square of distance. Thus, Newton inferred that "the gravity obeyed the inverse square law of distance." Afterward, instead of circular motion of the Moon in the case of which calculation was simple, Newton derived the gravity proportional to the inverse square of distance in the case of elliptic orbital motion of planet (Explanation 3.7).

Explanation 3.5 Deriving centrifugal acceleration $r\omega^2$

Let obtain the centrifugal acceleration in the case of the Moon on orbit around the Earth. Denote by r the distance between the Moon and the Earth (Fig. 3.13) and denote by θ the angle between the **r** and x-axis. The vector **r** expresses the position of the Moon. The position of the Moon is described as $\mathbf{r} = (x, y) = (r\cos\theta, r\sin\theta)$. The Moon receives the centrifugal force $\mathbf{F} = m\boldsymbol{\alpha}$ by the second law of motion where m is the mass of the Moon, $\boldsymbol{\alpha}$ is the vector of acceleration whose x-component is $\alpha_x = d^2x/dt^2$, and y-component is $\alpha_y = d^2y/dt^2$ (the second derivative with respect to time t). Using

$$x = r\cos\theta, \; y = r\sin\theta$$

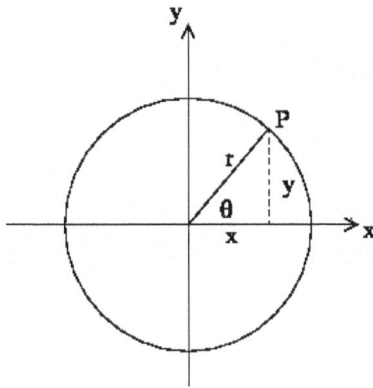

Fig. 3.13. Polar coordinate (r, θ).

Fig. 3.14.   Edmond Halley (1656–1742) (portrait in 1690).

Fig. 3.15.   Statue of Newton (taken by author in June 2016 at Trinity chapel).

and denoting by ω the derivative of θ with respect to time which is angler velocity and is constant, the absolute value of the acceleration is given by

$$\alpha = [(d^2x/dt^2)^2 + (d^2y/dt^2)^2]^{1/2}$$
$$= r[\{\cos\theta(d\theta/dt)^2 + \sin\theta(d^2\theta/dt^2)\}^2 + \{-\sin\theta(d\theta/dt)^2 + \cos\theta(d^2\theta/dt^2)\}^2]^{1/2}$$
$$= r[(d\theta/dt)^4 + (d^2\theta/dt^2)^2]^{1/2}$$
$$= r\omega^2$$

Hence the acceleration is the product of radius r multiplied by the square of angler velocity ω which is constant.

3.7.3 *Halley contributing to publishing Principia*

Halley (Edmond Halley) contributed to publishing *Principia* as follows. In January 1684, astronomer Halley who was interested in the orbital motion of planet, had a chance of discussion with Hooke on the motion of celestial bodies at a coffee house. Then Halley asked "Is the force influencing on planet revolving around the Sun, proportional to the inverse square of distance?" Hooke said "I thought to publish on the motion of planet after plenty of persons failed in solving the problem. It was the reason why I did not disclosed the detail of planet motion obeying the inverse square law of distance" (White, 1998:190, Gleick, 2003:124). However, when Halley asked the mathematical proof on the planet motion, Hooke's answer was obscure. Consequently, Halley doubted Hooke.

3.7.4 *Visiting Newton in Cambridge*

In Summer of the same year, Halley visited Newton in Cambridge. Halley inquired to Newton "If the gravity caused by the Sun obeys the inverse square law, then what is the orbital motion of planet?" Newton answered " It is an elliptic orbit," immediately. Newton promised "I calculated it a long time ago. Though I cannot give you the mathematical proof now, after recalculating it, I will send you the proof" (White, 1998:192). Newton who was not interested in mathematics and physics due to loneliness since Barrow passed away, devoted himself to treat mathematically the orbit of planet and the gravitation from that time. That is, the visit of Halley to Cambridge restored Newton's thinking mathematics and physics.

3.7.5 *Mathematical proof on orbits of planets*

Newton proved mathematically that the gravity proportional to the inverse square of distance, resulted in Kepler's second law which was law concerning area swept by linear line binding the Sun and planet. Furthermore, he indicated that if a planet moved in elliptical path under gravity, then the gravity was proportional to the inverse square of distance from the Sun. That is, if the gravity obeys the inverse square of distance, then a planet moves in elliptical path around an attracting body located at one focus of ellipse. Inversely, if orbit is an ellipse, then the gravity obeys the inverse square of distance. Newton completed manuscript on "mathematical proof on orbits of planets" under promise to Halley.

Explanation 3.6 Deriving ellipse from the inverse square law

Denote by M the mass of the Sun, by m the mass of a planet, by k the constant of universal gravitation, by $\mathbf{r}$ the position vector of the planet from the Sun.

Motion equation

On assuming the inverse square law, according to the second law of motion, we have the following motion equation:

$$m(d^2\mathbf{r}/dt^2) = -(kmM/r^2)\,(\mathbf{r}/r) \tag{1}$$

Vector product with **r** yields

$$\mathbf{r} \times m(d^2\mathbf{r}/dt^2) = -\mathbf{r} \times (kmM/r^2)\,(\mathbf{r}/r) = 0 \qquad (2)$$

because the absolute value of vector product of the vector A and vector B is the area of the parallelogram constructed by A and B and then $\mathbf{r} \times \mathbf{r} = 0$. On the other hand, angler momentum given by $\mathbf{r} \times m(d\mathbf{r}/dt)$ is constant that is proved by

$$d/dt(\mathbf{r} \times m(d\mathbf{r}/dt)) = \mathbf{r} \times m(d^2\mathbf{r}/dt^2) = 0 \qquad (3)$$

Hence $\mathbf{r} \times (d\mathbf{r}/dt)$ can be written by

$$\mathbf{r} \times (d\mathbf{r}/dt) = \text{constant} \qquad (4)$$

The velocity $(d\mathbf{r}/dt)$ is given by $\boldsymbol{\omega} \times \mathbf{r}$ where $\boldsymbol{\omega}$ is the angler velocity. Using the formula $A \times (B \times C) = (A \cdot C)B - (A \cdot B)C$ where $(A \cdot B)$ denotes a scalar product of A and B, we have

$$\mathbf{r} \times (d\mathbf{r}/dt) = \mathbf{r} \times (\boldsymbol{\omega} \times \mathbf{r}) = (\mathbf{r} \cdot \mathbf{r})\,\boldsymbol{\omega} - (\mathbf{r} \cdot \boldsymbol{\omega})\mathbf{r} = r^2\boldsymbol{\omega} \qquad (5)$$

because vector **r** is perpendicular to vector $\boldsymbol{\omega}$, then the scalar product $(\mathbf{r} \cdot \boldsymbol{\omega})$ is 0.

Kepler's second law

From (4) and (5), we have

$$r^2\boldsymbol{\omega} = r^2(d\theta/dt) = \text{constant} = C_1 \qquad (6)$$

where polar coordinate (r, θ) is defined in the plane containing orbit of planet as Fig. 3.16.

Equation (6) shows that the area velocity which is given by $r^2\boldsymbol{\omega}/2$, is constant. This is the Kepler's second law.

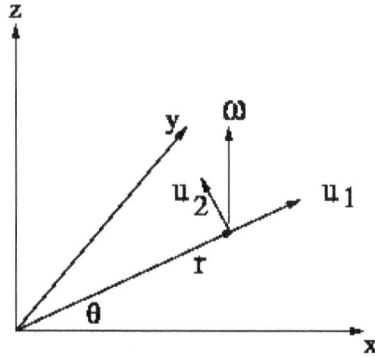

Fig. 3.16. The relation of three vectors of $\mathbf{u}_1$, $\mathbf{u}_2$, and $\boldsymbol{\omega}$.

The second derivative equation of r (scalar)

The acceleration vector $(d^2\mathbf{r}/dt^2)$ is expressed as

$$(d^2\mathbf{r}/dt^2) = \{(d^2r/dt^2) - r(d\theta/dt)^2\}\mathbf{u}_1 + \{2(dr/dt)(d\theta/dt) + r(d^2\theta/dt^2)\}\mathbf{u}_2 \quad (7)$$

where vectors $\mathbf{u}_1$ and $\mathbf{u}_2$ are defined as

$$\mathbf{u}_1 = (\cos\theta,\ \sin\theta),\ \mathbf{u}_2 = (-\sin\theta,\ \cos\theta) \quad (8)$$

$$\mathbf{r} = r\,\mathbf{u}_1,\ \boldsymbol{\omega}\times\mathbf{u}_1 = \mathbf{u}_2,\ \boldsymbol{\omega}\times\mathbf{u}_2 = -\mathbf{u}_1$$

$$d\mathbf{u}_1/dt = (d\theta/dt)\,\mathbf{u}_2,\ d\mathbf{u}_2/dt = -(d\theta/dt)\,\mathbf{u}_1$$

From (1) and (7), we have

$$\{(d^2r/dt^2) - r(d\theta/dt)^2 + (kM/r^2)\}\mathbf{u}_1$$
$$+ \{2(dr/dt)(d\theta/dt) + r(d^2\theta/dt^2)\}\mathbf{u}_2 = 0 \quad (9)$$

Because $\mathbf{u}_1$ and $\mathbf{u}_2$ are independent each other, their coefficients should be zero. Hence we have

$$(d^2r/dt^2) - r(d\theta/dt)^2 = -(kM/r^2) \quad (10)$$

$$2(dr/dt)(d\theta/dt) + r(d^2\theta/dt^2) = 0 \quad (11)$$

From (6) and (10), we get the second derivative equation of r

$$(d^2r/dt^2) - (C_1^2/r^3) = -(kM/r^2) \tag{12}$$

The second derivative equation of ρ

When we define ρ as

$$\rho = 1/r \tag{13}$$

using $d\rho/d\theta = (d\rho/dt)(dt/d\theta)$ we have

$$d\rho/d\theta = -(dr/dt)/C_1 \tag{14}$$

$$d^2\rho/d\theta^2 = -r^2(d^2r/dt^2)/C_1^2 \tag{15}$$

From (12) and (15), we have the second derivative equation of ρ with respect to θ:

$$d^2(\rho - \beta)/d\theta^2 = -(\rho - \beta), \quad \beta = kM/C_1^{\,2} \tag{16}$$

The solution of (16) is given by

$$\rho - \beta = \text{constant} \cdot \cos\theta$$

Hence we have the ellipse:

$$r = d/(1 - e\cos\theta) \tag{17}$$

where r in (9) – (17) is scalar, not vector, and

$d = C_1^{\,2}/(kM)$, $e = f/a$: eccentricity, f: focal length,
a: semimajor axis, b: semiminor axis,
using a and b, d is expressed as

$$d = b^2/a \tag{18}$$

Kepler's third law

The area S of ellipse which is orbit of a planet, is given by $S = \pi ab$, and the area velocity is $C_1/2$. Hence the period T of orbital movement of planet is given by

$$T = S/(C_1/2) = 2\pi ab/C_1$$

From (18), the ratio of the square of period to the cube of semimajor a is given by

$$T^2/a^3 = 4\pi^2/(kM) \tag{19}$$

Hence T^2/a^3 is constant irrespective of any planet. This is the Kepler's third law.

Explanation 3.7 Deriving the inverse square law from elliptic orbit

We use the same variables and symbols as Explanation 3.6. First we consider the derivative equation of ρ with respect to θ whose solution is an ellipse. Then indicate that the Kepler's third law derives the Kepler's second law. We express the derivative equation in the form of derivative equation of **r** with respect to time. From this equation and the Kepler's second law, we can obtain the second derivative of **r** with respect to time and derive the inverse square law.

3.7.6 *Completion of manuscript for Principia*

In Autumn 1684, Newton handed the paper of 9 pages "On motion of a body revolving (De Motu Corporum in Gyrum) over to Halley by mathematician Paget (Edward Paget) who was a mutual academic acquaintance. This paper explained the theoretical concept on centrifugal force influencing a body revolving, and was the basis of Book I in *Principia* published 3 years later.

In Spring 1686, the manuscript of *Principia* almost completed, and it was composed of Introduction, Books I, II and III though it was not exactly perfect. Books I and II treated force and motion. Books III described the application of theoretical concepts in Books I and II. Newton's motion's three laws were described in Introduction. The complete manuscript of Book III was brought to Halley on 4th April 1687.

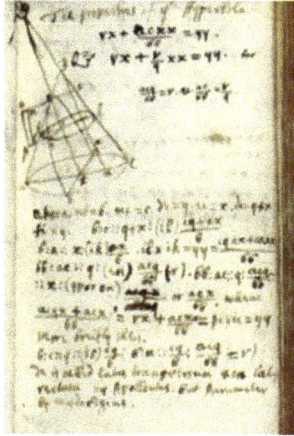

Fig. 3.17.   Newton's manuscript given by Dr. Matsuda in Sep. 2023.

PHILOSOPHIÆ

NATURALIS

PRINCIPIA

MATHEMATICA·

Autore *J S. NEWTON,* *Trin. Coll. Cantab. Soc.* Mathefeos Profeffore *Lucafiano,* & Societatis Regalis Sodali.

IMPRIMATUR·
S. PEPYS, *Reg. Soc.* PRÆSES.
Julii 5. 1686.

LONDINI,

Juffu *Societatis Regiæ* ac Typis *Jofephi Streater.* Proftat apud plures Bibliopolas. *Anno* MDCLXXXVII.

Fig. 3.18.   First version of *Principia* (language: Latin)

Principia was written in the form of propositions as unreadable as possible deliberately, in order to avoid being bated by person having a smattering of mathematics. That is, it was written in the form such that the previous proposition had to be understood for understanding a proposition.

Under the information by Humphrey (Humphrey Newton) who served as assistant and transcribed Newton's note, Newton forgot eating and very rarely went to bed till 2 or 3 of the clock devoting himself to the research during completing manuscript of *Principia* (White, 1998:213).

3.7.7 *Principia published by Halley's own money*

In May 1685, before receiving complete manuscript, Halley got approval of plan publishing manuscript by Newton at meeting of the Royal Society. Because the finance of the Royal Society was then bankrupt, Halley used his own money for publishing the manuscript. In July 1687, *Principia* was published, and was evaluated as the greatest scientific book. In this book, Newton indicated the principle which elucidated "motion of celestial body in universe." The principle became the doctrine in new epoch. He applied the principle to analyzing motion of body under gravity, that is, orbital motion, motion of thrown body, pendulum, and free falling body.

Furthermore, he explained the law of universal gravitation, that is, "all bodies in universe mutually influenced gravitation whose magnitude was proportional to product of mutual masses, and was proportional to the inverse square of distance." Precession of the equinoxes of the Earth, and motion of the Moon receiving perturbation by the gravitation of the Sun were explained.

3.7.8 *Remote action of force: Universal gravitation*

Principle described in *Principia* was so high level that Newton was evaluated as the leader in science. However, scientists in the Continent did not accept the idea "remote action of force" that insists force leaping to far away body, that is, Newton's idea of universal gravitation. The scientists in the Continent accepted Descartes' idea "proximity action of force," that is, "force influencing celestial bodies was transmitted in close proximity through aethel which fulfilled all space in cosmos and was invisible and massless. The motions of celestial bodies were caused by vortices of aethel." Descartes' idea was easy to understand for them. Thus, because they believed the idea of "proximity action of force" by Descartes, they did not agree the idea "remote action of force" by Newton. But, counter-argument like this could not apply the brakes to worldwide praise for Newton's great achievement.

It was estimated that Newton was influenced by the idea of alchemy, over which he was enthusiastic, that is, "force was based on mysterious action." This was the reason why he rejected Descartes's idea.

3.8 Emergency of University

3.8.1 *Intervention of James II*

After Charles II passed away of disease, stubborn and unpopular James II took the throne. Because the King was Catholic, he had thoughtless plan to let all English be Catholic. Then, in Great Britain there were Catholic and Protestant, and they lived calmly without antagonism. The King's plan was meaningless and was not supported by people.

Fig. 3.19. Newton (portrait in 1689).

Cambridge University was Protestant's fortress. In February 1687, the King imposed unreasonable demand. James II ordered the university to install a Benedictine monk as a Master of Arts, with an exemption from the required examination and oaths to Anglican Church (Gleick, 2003:146). A Master of Arts had a right to vote in board of directors, and had a voice about management of university. Therefore, if university obeyed the orders then person convenient for the King was sent, and university might have fear to be intervened.

3.8.2 *University's firm attitude according to Newton's advise*

Devout Protestant Newton suspected the emergency, and advised vice-Chancellor of university, that university should not submit to tyrannical order. The vice-Chancellor made a document expressing firm attitude according to Newton's advise. The King lost his temper for attitude of university, and ordered that eight representatives should report themselves to a Commission for Ecclesiastical Causes.

The board of directors elected Newton among eight representatives. Before leaving for London, eight representatives discussed measures. Though in order to prevent from being opposed to the King, compromise was proposed, Newton insisted that they should attend with firm attitude because if they were at the mercy of the King, then Catholics would be sent successively.

At the Commission for Ecclesiastical Causes, the King side exercised their authority but university side was not overcome with fear, and did not bow down to the King. Consequently, university side did not obey the King, and could defend autonomy and freedom of scholarship. Thus, university crushed the thoughtless plan of the King.

At Oxford University which was a Protestant fortress as Cambridge University, students defiant to the King's attitude rebelled against the King, and the King sent equestrian soldiers. By that time, James II lost support of people.

By influential leader, under the maneuver changing King peacefully, William III was sounded out on the next King. In 1688, William III (Willem III van Oranje-Nassau) arrived at England taking the lead of

Netherlandish fleet, and then James II escaped to France thinking his disadvantage.

In 1689, Cambridge University elected Newton as one of two Members of Parliament. The reason was that Newton had firm attitude to the thoughtless intervention of James II, and was the leader saving the emergency of university. In the same year at Parliament James II abdicated, and William III succeeded the throne.

When as the Member of Parliament Newton stayed in London, he felt the pleasant life in the city different from country. It caused that he would wish to get a post in London.

3.9 Life in London

3.9.1 *Exhaustion*

In September 1693, he suffered from lack of sleep and poor appetite. The reason was that the mental was exhausted due to extraordinary concentration in writing and due to the anxiety about continuity of creative energy which he felt after the praise for *Principia*.

Newton thought that vigorous life in London could save the exhaustion and fear. He desired his friend in London to get a post in London.

Fig. 3.20. Newton (portrait in 1702).

However, he could not get reply fulfilling his desire, and his mental exhaustion more and more increased. When in order to divert his attention from the exhaustion, he researched in experimental room, he discovered the "cooling law" that is, "thermal loss between two bodies is proportional to the difference of temperature." Furthermore, he found that boiling and melting occurred at constant temperatures.

3.9.2 *Master of the Mint*

Newton who was more and more exhausted due to lack of sleep, sent again to friend a letter desiring to get a post in London. In 1695, Newton got a post of Warden of the Mint. In 1699, he became Master of the Mint. His work there was recoinage.

Then circulating silver coins were cut. People thought that the value of silver coins like this were low. The government planned to cast new silver coins. However, on new silver coins circulating, people would store up new silver coins thinking the value of new coin being high, and only old coins would circulate. Therefore the effect of recoinage would not be expected. That is, it was estimated the economical situation "bad money drove out good money" expressed in Gresham's law.

Consequently, Newton, etc. concluded to scrap all old cut coins and to cast new coins. To execute the plan, for preventing from confusion, it was necessary to cast vast amount of silver coins as short time as possible in the Mint. It might be estimated that civil servant would resist sudden hard work. However, Newton decided to devote himself to the nation, and be engaged in difficult enterprise of recoinage. Analyzing the work process and improving the efficiency of the process, he was successful in the enterprise. He sought counterfeiters and prosecuted them.

Till he resigned Professor at Cambridge in 1701, concurrently as Master of the Mint he worked with enthusiasm. He was satisfied with his life in London, and in life he held the position of Master of the Mint. He summoned his favorite niece to London. She was sociable and had different personality from uncle Newton. She was witty beautiful woman, and took care of her uncle, and played a role in social circle.

3.9.3 *Controversy with Leibniz*

Though Newton lived a calm life, unexpected controversy with Leibniz occurred. Leibniz insisted that he invented differentiation and integration calculus, and Newton never invented it. Leibniz wrote manuscript on differentiation and integration calculus in 1675 and in 1686 he published it. On the other hand, in 1666 when university was closed, Newton invented differentiation and integration calculus. Friend of Newton knew Newton's works on invention of differentiation and integration calculus. Today, it is thought that both scientists independently invented the calculus.

In November 1703, the Royal Society elected Newton as President for his great work. In February 1704, Newton presented the book *Optics* by him to the Royal Society. It was not written in Latin but in English. Because of Hooke's criticism on the paper "Theory of Light and Colors" carried in 1672, the book was not published immediately by Newton's intention, and in 1703 after Hooke passed away, the book appeared broking the silence at about 30 years after the paper had been carried. The book described refraction of light, reflection, rainbow, mirror and prism's function on the basis of a series of his experiments. In 1705, Newton who was the worldwide famous scientist as the author of *Principia*, was knighted by Queen Anne. Newton was the first scientist given this honor.

3.9.4 *Illness*

At the age of 81, Newton was directed by doctor to concentrate on cure because symptoms of kidney appeared. Afterward, pneumonia was complicated. Nice Catherine was astonished by serious illness of uncle, and devoted herself to taking care of uncle. In order to prevent from air pollution in London, she intended to make uncle recuperate under a change of air in Kensington.

Despite illness, Newton continued work on revision of *Principia*, and researched on theology. He attended the regular meeting on every week. On 2nd March 1727, feeling cold he went out to attend the meeting of the Royal Society. For him with pneumonia complicated, the going out was great burden, and at going back home his illness became so serious that doctor could not cure. Consequently, he should be in the bed. Afterward, during two weeks, a coma and consciousness appeared alternately. On consciousness, he talked to a collegue (John Conduitt) of the Mint and in-law nephew with smile "I have not intention accepting the final ceremony as Protestant" (White, 1998:360). On 20th March 1727, at the age of 84, genius Newton passed away.

Newton enrolled at Cambridge University as a sizars. Afterward, he exhibited gift of genius, and accomplished great discoveries on physics, astronomy, and optics, and accomplished invention on mathematics. By *Principia*, motion of celestial body was elucidated. *Principia* became the doctrine of science in new epoch. Newton who noted down that "Truth is my greater friend" in his youth, was buried as Sir Isaac Newton in Westminster Abbey. The inheritor constructed his monument which watched the corner called as "Scientist Corner" where famous English scientists such as Darwin (Charles Darwin) and Maxwell were buried.

Appendix 3.2 After Newtonian mechanics

Rigid body mechanics

Motion's laws of Newton were concerning dimensionless mass point. However, real body has dimension, and the motion of such body should be derived from the motion of mass point. The person who played important role on motion of body with volume, was d'Alembert (Jean Le Rond d'Alembert) (Yukawa & Tamura, 1955–1962:I). In 1743, he founded mechanics of rigid body. Rigid body is defined as body with volume never transforming (Explanation 3.8).

Analytical mechanics

New mathematical analysis for kinetic theory made progress, and rich mathematical methods contributed to development of mechanics. Great mathematicians in the 18th century Euler (Leonhard Euler) and Lagrange (Joseph Louis Lagrange) related "mechanics of rigid body" to Newtonian mechanics, and founded "Analytical mechanics" which was mediator between Newtonian mechanics and quantum mechanics.

Electromagnetic theory

In the 19th century, electromagnetic phenomena were researched by Faraday who proposed electric and magnetic fields, that is, force field (Appendix 4.3). Based on Faraday's force field, Maxwell was successful to derive electromagnetic equations consolidating electromagnetic phenomena. He predicted theoretically the existence of electromagnetic wave and predicted that light was identical to electromagnetic wave. The prediction by Maxwell was verified by Heltz (Appendix 3.1). Afterward the electromagnetic wave would indicate the application limit of Newtonian mechanics.

The end of the 19th century

At the end of the 19th century, by discovery of X-ray, cathode ray and radioactive ray, physical field proceeded toward microscopic field such as atom. In the microscopic field, assuming that Newtonian mechanics held true, statistic consideration by Maxwell and Boltzmann was useful to explain the statistic, thermo-dynamical properties.

However, the problem on blackbody radiation could not be explained by classical physics. This heat radiation was electromagnetic wave with long wave length, and invisible infrared light. Intensity of heat radiation changed depending on frequency with peak at some frequency. The peak shifts to high frequency for high temperature of black body on the basis of Wien's Displacement Law. Theoretical explanation for the experimental data were tried. However, the estimation by the classical theory was not at all coincide the experimental data. The try of theoretical explanation of blackbody radiation had failed.

Energy quantum

In 1900, Planck derived theoretical formula coinciding experimental data. This was Plank's formula (Appendix 6.2). He was aware of the important quantity called "energy quantum" (Appendix 3.1). Energy of light is given by energy quantum as a unit, multiplied by non-negative integer. That is, he discovered that energy was discrete quantity. Planck's discovery played the important role of starting quantum mechanics following Newtonian mechanics.

The special relativistic theory and quantum mechanics

The Michelson–Moley experiment indicated the necessity of reconsidering time-space concept, resulting in the special relativistic theory (Paragraph 6.6). On the other hand, in microscopic field, it was inevitable to generalize Newtonian mechanics fundamentally, resulting in quantum mechanics as a new mechanics. Thus, at the beginning of the 20th century, application limit of Newtonian mechanics was indicated, and Newtonian mechanics held place as classical physics.

Explanation 3.8 Rigid body mechanics

As Fig. 3.21 rigid body is a system of mass points in which distance between mass points is invariant.

The state of mass point is expressed by indicating the position in three-dimensional space. On the other hand, the state of rigid body is expressed by indicating not only the position in 3 dimensional space but also rotational axis and angle around the axis because of rigid body's volume.

The motion equation is expressed as follows:

[all mass] × [acceleration of center of gravity] = [sum of forces]   (1)

Temporal differentiation of [sum of angular momentum
of mass point] = [sum of moments of forces]   (2)

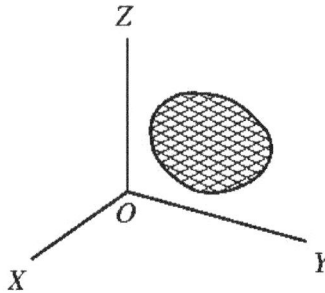

Fig. 3.21.   Rigid body.

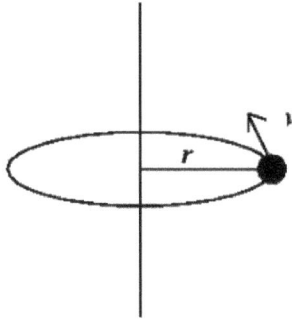

Fig. 3.22. Revolution around axis.

Tough in Newtonian mechanics, motion equation is expressed by indicating equation of acceleration of mass point, in rigid body, motion equation is expressed by indicating not only equation of center of gravity of rigid body but also equation (2) expressing rotation of rigid body because of volume of rigid body. Equations (1) and (2) decide the position and posture of rigid body. Angular momentum is proportional to product of velocity multiplied by distance between mass point and rotation axis. Moment of force is proportional to product of force by distance from rotation axis.

For example, as Fig. 3.22, when force is given to mass point at position of radius r revolving with velocity v around axis, equation (2) indicates that temporal change of angular momentum is equal to moment of force. In case of no moment of force, right-hand side of equation (2) is zero, and angular momentum which is proportional to product of radius by velocity, is constant. Therefore in case of no moment of force, as Fig. 3.22, decreasing radius increases the velocity v. Because velocity is given by product of radius multiplied by angular velocity, the above mention is the reason why on figure skating with spin, bringing spread arms close to trunk as axis increases angular velocity.

Explanation 3.9 Minimum action principle

Minimum action principle is a principle in physics using variational method. We consider the moving path of mass point from time t_1 to time

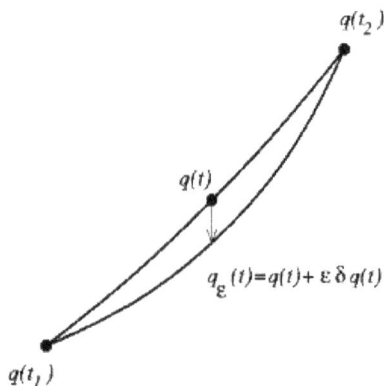

Fig. 3.23.   Variation $\varepsilon\delta q(t)$ of the moving path.

t_2 in a force field. We define Lagrangean L as difference between kinetic energy and potential energy. For example, when a mass point is placed in field of gravity on the ground, mass point has the larger potential energy for the higher position. Kinetic energy is proportional to the square of velocity of mass point. We define action integral as the integral of L from time t_1 to t_2. Minimum action principle gives the equation deciding the moving path of mass point with the minimum action integral. This equation is called Euler equation.

We define the path $q(t)$ as the position at time t between t_1 and t_2. As figure it is assumed that a new path $q_\varepsilon(t)$ is given by variation of path $\varepsilon\delta q(t)$. The variation is assumed to be zero at end points $q(t_1)$ and $q(t_2)$. ε is assumed to be small.

Euler equation is derived from the necessary condition that if the path $q(t)$ has the minimum action integral, the variation of action integral should be zero for any variation of path.

In the case where the mass point with mass m is moving in the field of gravity of the Earth, then we have the Lagrangean L:

$$L = T - U \qquad (1)$$

where $T = (1/2)m(dq/dt)^2$ (kinetic energy), U: potential energy. The mass point receives the force $f(q, t)$:

$$f(q, t) = -\partial U/\partial q \qquad (2)$$

The Euler equation is given by

$$d/dt[\partial L/\partial(dq/dt)] = \partial L/\partial q \qquad (3)$$

From (1), we have

$$d/dt(\partial L/\partial(dq/dt)) = m(d^2q/dt^2) \qquad (4)$$

$$\partial L/\partial q = -\partial U/\partial q = f(q, t) \qquad (5)$$

Hence the Euler equation (3) decides that the mass point moves according to the second law of motion:

$$m(d^2q/dt^2) = f(q, t) \qquad (6)$$

That is, Newton's motion equation is expressed by using minimum action principle. Using minimum action principle, Euler and Lagrange expressed Newtonian mechanics with the mathematically excellent form (Yukawa & Tamura 1955–1962:I).

Chapter 4

Michael Faraday

Michael Faraday (1791–1867). Sketch by author.

> Michael Faraday discovered electromagnetic induction which caused inventions of generator, motor and transformer by which mankind life was improved rapidly. He researched plenty of electromagnetic phenomena such as electromagnetic induction, dielectric polarization of dielectrics, magnetization of magnetic substance and magneto-optical effect: Faraday's effect. Electromagnetic theory was founded by Maxwell based on experimental researches by Faraday. Electromagnetic theory and Newtonian mechanics constituted the greatest two theories of classical physics until the end of the 19th century.

4.1 Upbringing

4.1.1 *Birth of Faraday*

Faraday's father (James Faraday) was a blacksmith working for the ironmonger Boyd (James Boyd). In 1787, he and his wife (Margaret) relocated to Butts by Boyd's recommendation. On 26th May in the same year, Faraday's sister (Elizabeth) was born, and on 8th October in the next year, brother (Robert) was born. 3 years later, on 22nd September 1791, Faraday was born in Newington Butts which was then at the edge of London: beyond lay the fields of Surrey. (Butts is included in London at present). Faraday was named after his maternal grandfather (Michael Hastwell).

When Faraday was at the age of 5, the family relocated to rooms over a coach house in Jacob's Mews in west London near Welbeck Street where his father worked. His father, in his mid-40s, suffered from so ill health that he could not work during long time per day. Therefore, it was difficult to get enough income for feeding the family. In 1801, the price of bread peaked because of shortages occasioned by the war with France (James, 2010:10), and they were badly off. Consequently, the family received public relief. However, at the beginning of the 19th century in Great Britain public facilities such as a workhouse treated poor people coarsely. The family of Faraday might endure the insolent attitude for people suffering from poverty. In fact, Faraday was given only a loaf of bread per week. His mother took in lodgers for a living.

His education was the most ordinary description, consisting of little more than the rudiments of reading, writing, and arithmetic at a common day-school. His hours out of school were passed at home and in the streets (James, 1991:xxvii).

Then social background was as follows. In Great Britain, the industrial revolution which occurred at Manchester and Liver-pool in Lancashire, was in progress. Under King George III, London was the largest, richest, and most powerful city in the world.

On the other hand, in the Continent, French Revolution beginning in 1789, was in progress. In 1793, monarchy was abolished, and Louis XVI and Queen Marie-Antoinette were executed. France after the Revolution was opposed to Great Britain. Napoleon Bonaparte who became Emperor, antagonized Great Britain. In 1837, the Victorian age commenced in Great Britain.

4.1.2 *Apprenticeship*

To assist his poverty-stricken family, Faraday at the age of 13 should work. On 22nd September 1804 a birthday, he was employed as a newspaper-cumerrand boy at bookshop and bookseller Frenchman Riebau (George Riebau) at 2 Blandford Street, near his address. His work was to deliver the bound books and newspapers · magazines. To collect magazine was also his work. Because the charge of subscribing to magazine was expensive, most people returned it, that is, rented magazine. Faraday was busy with work every day.

Due to his experience, afterward when he met newspaper cumerrand boy, always he kindly talked to him. He said his niece "Because I had experience of carrying newspaper, on meeting him I am kind for him" (James, 2010:20).

Riebau was a bookbinding artisan at first class. Generally, person working under such artisan became apprentice for learning bookbinding. To be apprentice, fee as premium was necessary. But Faraday could not pay the fee. Previously his brother was apprenticed to blacksmith, and his father could not afford to pay for apprenticeship of Faraday. However, Riebau loved Faraday of 14 years of age as his own son, and permitted for

Faraday to be apprenticed to Riebau. On 7th October 1805, the indenture for apprenticeship was made. The indenture described the terms that instead of no premium, he should faithfully work, and not go in and out taverns. After completing apprenticeship of 7 years, apprentice became artisan and got wages. He was very skillful, and learned excellently bookbinding.

Faraday took delight in reading various scientific book (including Jane Marcet's Conversations on Chemistry and the electrical treatise by James Tytler in the third edition of the *Encyclopaedia Britanica*) (James, 1991:xxix) bound beautifully by him. Learning scientific knowledge by reading books, he was interested in science. For Faraday with the only elementary education, book was the only teacher. He did chemical experiment to confirm the knowledge obtained from book, using free time in home after work. He fabricated the device causing friction electricity which is now held in the Royal Institution.

4.1.3 *Lecture by Tatum*

From 19th February 1810 to 26th September 1811 during apprenticeship, Faraday attended the lecture by Tatum (John Tatum) the leader of the City Philosophical Society in London. This society had been founded in 1808 to help give artisans and apprentices, like Faraday, additional access to scientific knowledge. There were many such lectures and societies in London and provincial cities during the first quarter of the 19th century. Those who attended such societies believed that self-improvement would lead to better prospects for themselves both materially and morally. Materially because it would continue to help with the ever increasing pace of the industrialization of Britain and morally because as Faraday himself put it, philosophic men had superior moral feelings (James, 1991:xxix). The lecture by Tatum consisted of 12 lectures. Among lectures, seven lectures were concerning to electricity. His brother who previously became blacksmith and understood always for Faraday to be interested in science, paid 12 shilling for fee of 12 lectures. Then, his brother took care of Faraday's family. On 30th October 1810, his father passed away.

The lecture by Tatum explained the advanced scientific knowledge. Experiments on Leiden's bottle and electro-chemical decomposition of salts were shown. This lecture provided him the knowledges equivalent to contents in higher education. He obtained wide and higher scientific knowledge.

When he could not understand the contents in lecture, on the same day, he consulted encyclopedia Britannica to make up his own knowledges. He was unusually eager to study knowledges. He completely understood all the lectures, and completed four tidy notebooks. This notebooks indicated how Faraday who could not receive regular schooling due to poverty, yearned scholarship.

Attending the lecture, he got many true friends. Among the friends, there was Phillips (Richard Phillips). He who formed life-long friendships with Faraday, afterward became an excellent chemist.

4.2 Davy, Professor of Chemistry in the Royal Institution

4.2.1 *Lecture by Davy*

Among the customers at bookshop of Riebau, there was Dance (the son of William Dance) who was a subscriber and Life Member of the Royal Institution. The encounter of Faraday with Dance led to turning point in his life.

Dance who had good will to Faraday working seriously, had opportunity to read excellently tidy notebook by Faraday. The notebook indicated that Faraday understood contents of all lectures completely. Dance was astonished that Faraday receiving no higher education, so understood scientific knowledge, and thought that Faraday had extraordinary brain.

In 1812, Dance presented a ticket to attend the lectures by Davy (Humphry Davy) Professor of Chemistry in the Royal Institution, to Faraday of 21 years of age. In December 1807, the Academie des Sciences awarded Davy the Volta prize of 3000 francs founded by Napoleon antagonizing Great Britain. Davy was the most outstanding scientist in Europe.

Lecture by Davy was delivered with experiments such as electro-chemical decomposition. Then, acids was considered to contain oxygen. Davy indicated that muriatic acid was a compound of hydrogen and chlorine and contain no oxygen. The experimental equipment which Faraday knew only in a book, was excellently produced by professional. Faraday marveled at such equipment by professional.

Fig. 4.1.   Humphry Davy (1778–1829) (portrait in 1803).

Faraday precisely summarized the lecture in his notebook. Davy's lecture was interesting for him. The lecture was the final lecture of Professor of Chemistry in the Royal Institution. On 8th April 1812, Davy was knighted by the Prince Regent (James, 2010:31), and three days later, a very wealthy widow (Jane Apreece) became his wife. Because of no need to work due to her wealth, at the age of 34, he retired from being the Royal Institution's Professor of Chemistry. But the Royal Institution's manager was anxious to retain their connection with the most famous English chemist of the time, and appointed Davy Honorary Professor of Chemistry and Experimental chief. Consequently, Davy had great influence in the Royal Institution after marriage.

Explanation 4.1 The Royal Institution of Great Britain

In 1799, at 21 Albemarle Street, the Royal Institution was founded. The purpose was "useful mechanical inventions and improvement, the application of science to the common purpose of life, and diffusing the knowledge by courses of philosophical lectures and experiments (James, 2010:27). Subscribers on foundation became Life Members.

Then, at colliery, explosion often occurred. Explosion occurred because of fire-damp in colliery catching fire by miner's lamp. By safety

Fig. 4.2. The Royal Institution (painting in 1838).

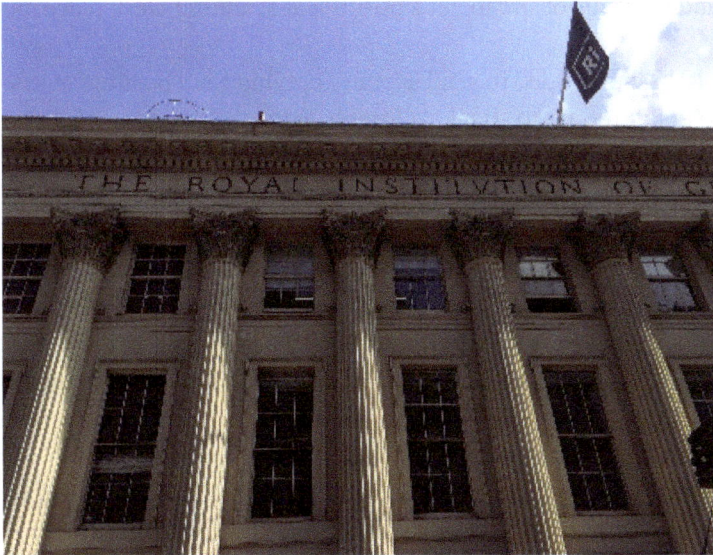

Fig. 4.3. The present Royal Institution (taken by Dr. Matsuda in Sep. 2017).

lamp invented by Davy plenty of worker's lives were saved. To begin with Davy and Faraday, plenty of scientists such as Dewar (James Dewar) who was pioneer of low temperature science, and succeeded liquefaction of hydrogen gas, and invented Dewar bottle which preserved liquefied gas keeping extreme low temperature, and Bragg's father and son (William Henry Bragg 1862–1942, William Lawrence Bragg 1890–1971) researched in the Royal Institution. Thus, during about 200 years in the Royal Institution, plenty of geniuses had researched, contributing to development of science. Since 2007, it was not used as the place of research. Since 1973, Faraday Museum opened in the Royal Institution.

4.2.2 *Yearning for scientist*

Since attending Davy's lecture, Faraday yearned to be a scientific practitioner. For him the Royal Institution was the most fascinating work place.

In October 1812 at the age of 21, his apprenticeship ended, and he started his journeyman career with De La Roche. New proprietor was whimsical, and not a person who encouraged Faraday studying with enthusiasm. Because of inferior employment for employee, the return home became late, and experiment using free time was difficult. Then, more and more his desire to get a position at scientific work place, became strong.

In 1812, he decided to send a letter applying for a position at scientific experimental room, to Banks (Joseph Banks) President of the Royal Society. He guardedly completed the letter asking to be "engaged in scientific occupation, even though of the lowest kind" (James, 1991:xxx). By the memory of delivering books at Riebau's bookshop, he went Banks' residence, and asked gatekeeper to pass on the letter, and said that he would come for receiving the reply. According to promise, one week later he went to the residence, and received just his own envelope sent one week ago, on which "no reply" was scribbled. Afterward, he sought a job but always was rejected. The reason was neither academic career nor qualification. Anyone did not inquire what he knew and could do. Faraday thought that a person with neither academic career nor qualification could not get a post at scientific workplace, and he lost confidence (Sootin, 1976:35).

The number of scientific practitioners in Great Britain was small then. Most practitioners were earning their living in other professions or possessing private wealth. Under social background like this, it was difficult for Faraday without private wealth to seek a job.

Since he relocated to De La Roche, he often visited bookshop of Riebau who was very kind and helpful to Faraday. One day when he visited the bookshop of Riebau, Dance happened to be there. Then, Dance suggested for Faraday to send a letter applying for a post in the Royal Institution directly to Davy, attaching his notebook summarizing the lecture by Davy. But Faraday talked that he sent a letter to Banks but he received no reply. However, Dance persuaded for him to try his best tenaciously. Consequently, Faraday followed the persuasion.

Davy who received Faraday's letter and notebook, showed the letter to a director Pepys (John Pepys), and sought his advice talking "a youth called Faraday asks to be employed in the Royal Institution. What can I do?" Pepys answered "Let him wash test tubes. If he is a valuable person, he will wash obediently test tubes. If he refuse it, he is not worth supporting." To this answer, Davy said "I cannot do such examination. I desire to judge him with better method" (Tyndall, 2002:2). Davy then was Honorary Professor of Chemistry in the Royal Institution and the head of Laboratory.

At Christmas Eve in 1812, one week after Faraday posted the letter and notebook, the reply from Davy was delivered to him. It was a letter understanding Faraday's serious attitude for science. In the letter, Davy notified that he would leave London, and would meet on returning in January 1813.

4.2.3 *Interview with Davy*

At the beginning in the next year, he was informed about the time and date following the promise. It was arranged that Faraday would meet with Davy in the Royal Institution. Faraday went to the Royal Institution with feeling of expectation mingling with anxiety.

On interview Davy inquired why Faraday wished to be a scientific practitioner. Faraday explained his academic enthusiasm. Davy indicated the severity of science as "Science is harsh mistress, and serving it is poor paid" (James, 1991:497). However, Faraday answered that scientists obtained great reward from seeking the truth by researching science, and that science let human have superior moral feeling, released from worldly low level thought (Sootin, 1976:51).

Davy inquired whether Faraday's knowledge concerning science was obtained by self-study. Faraday who received almost no regular education, thought that he would be inquired about academic career, and got nervous remembering previous sad experience in seeking a job. However, Davy did not inquired about academic career, and said that the tidy notebook summarizing lecture by Davy, indicated Faraday's enthusiasm for science, good memory and capacity (Sootin, 1976:54).

Davy's career in youth was similar to Faraday's career. He during his apprenticeship of apothecary, read *Traite Elementaire de Chemie* by Lavoisier (Antoine-Laurent de Lavoisier) and took a strong interest in chemistry. His memory of hardship at boyhood overlapped to Faraday's hardship. Davy was very kind to Faraday.

At final of interview, Davy advised "It is the most reliable to continue the bookbinding artisan," and notified that there was no vacant post. However, when parting, Davy said that when a vacant post would occur, he would remember Faraday (Sootin, 1976:55). The Davy's last word gave a hope to Faraday. In fact, this word would afterward change Faraday's life.

4.2.4 *3 days' work at the Royal Institution*

Two days after the interview, a letter from Davy was received. Davy at this time was suffering from injuries sustained following an explosion caused by combining nitrogen and chlorine. Glass had penetrated his eyes, and impairment of vision was continued. Because the dead line for submitting paper got close, the letter informed the desire that Faraday went to the Royal Institution for making a fair copy of draft of experimental note by Davy during 3 days.

Faraday could easily understood all academic terms described in experimental note due to his endeavor, and could accomplished the 3 days work. Davy was satisfied for Faraday's work because Faraday had ability for understanding scientific knowledges, and his handwriting was very beautiful. For Davy, Faraday became a person as favorite existence.

The 3 days works were those giving opportunity to Faraday. If Faraday could not understand academic terms, and could not make a fair copy of draft, then the future prospect in his life would not open. If there is no ability catching the opportunity getting close, the opportunity immediately gets away. He could get excellently the opportunity.

One night in February after he made a fair copy of draft, on undressing for bed, a splendid coach pulled up at Faraday's home 8 Weimous Street where Faraday's family lived since 1809. Davy's footman delivered a note requesting Faraday to call the following morning (James, 2010:34).

The Davy's assistant Payne (William Payne) gave a trouble of beating Newman (John Newman) delivering experimental apparatus to the Royal Institution, and immediately was dismissed. Because of vacant post, the employing Faraday was informed. Davy kept his promise at the interview with Faraday.

4.3 Opening the Doorway to Researcher

4.3.1 *Journey for the continent*

On 1st March 1813, Faraday served as a experimental assistant in the Royal Institution. Davy Honorary Professor had great dignity, and utilized Faraday employed in the Royal Institution for his private object.

On 13th October, 7 months after he served in the Royal Institution, Davy departed with Faraday for the Continent. For Faraday, this journey was the opportunity to get extensive knowledge. They practiced demonstration of chemical experiment and interviewed at various countries. In Paris as the first journey, Davy showed the property of iodine before chemists in France. In his correspondence, Faraday described "Iodine is heavy black element like lead. Heating it, it melts and becomes beautiful purple gas. Cooling it, it becomes crystal. Iodine forms compound with all metal except platinum and gold" (James, 1991:74). Davy submitted paper summarizing experimental results on iodine to the Royal Society. They interviewed Ampere (Andre Marie Ampere). Ampere discovered Ampere's law concerning electric current 7 years after the interview.

Through Lyons, Nice from Paris, they went across Alps of 6000 feet France-Italy border, and went Turin, Genoa and Florence.

Fig. 4.4.   Andre-Marie Ampere (1775–1836) (portrait in 1825).

4.3.2 *In Florence*

In Florence, Davy practiced combustion experiment of diamond placed at focus of lens. Diamond was placed in the middle of a glass globe supported in a cradle of platinum. To continue combustion, glass globe was pierced full of holes. Diamond placed at the focus of lens was heated, and burned with beautiful vivid scarlet light. Afterward, in glass globe it was found to contain nothing but a mixture of Carbonic and Oxygen gases. This proved that the diamond was pure crystalized carbon.

In correspondence, Faraday said about Florence "Florence is beautiful, and preserves enormous amount of instructive things, and is city like fine Museum of Natural History. The telescope with which Galileo discovered the satellites of the Jupiter (Fig. 1.22), and lens polished at the first time by Galileo are left by Galileo" (James, 1991:75).

They went to Rome, Naples and Milan. In Milan, they interviewed old Volta (Alessandro Volta) who invented electric battery. In Geneva Faraday

Fig. 4.5. Ponte Vecchio painted by Antonietta Brandeis (1848–1926).

Fig. 4.6. Jean-Baptiste Andre Dumas (1800–1884).(portrait in 1840–1850).

made friends with Dumas (Jean-Baptiste Andre Dumas) 15 years younger than him. Dumas afterward contributed to organic chemistry and became Professor of Chemistry in Ecole Polytechnique.

They went from Geneva to Lausanne, Berne, Zurich, Munich, and went across the Tyrol Alps Austria-Italy border, and back to Italy again, and went to Padua and Venice.

4.3.3 *Napoleon's escape from Elba*

Through Bologna and Florence, they arrived Naples. Their schedule was to go to Constantinople. But hearing Napoleon's escape from Elba, they decided to end their journey, being afraid of unstable political situation. Going across again the Tyrol Alps, they went Germany, Netherlands and Belgium, and sailed across to Great Britain. Finishing long journey of one year and 6 months, on 23rd April 1815, they arrived London. Immediately after, "Battle of Waterloo" of French forces commanded by Napoleon vs Britain-Holland-Prussia allied forces occurred in Belgium in June.

Explanation 4.2 Davy lamp

On research of miner's safety lamp, Davy verified that in case of inner radius of metal tube less than 0.1428 inch, and the depth proportionally long to inner radius, mixture of air and flammable gas did not explode, and explosion could not go through tube. Heat loss on the surface of metal with thermal conduction contributed to this phenomena. On the basis of the above results he fabricated safety lamp as follows. For taking light side planes was constituted with 4 glasses, upper and bottom surfaces were constituted with metals. To continue combustion, bottom metal was pierced to have small holes with diameter of 0.125 inch and depth of 1.5 inch. Upper surface metal was pierced to have small holes for vent playing role of chimney. Thus, inside and outside of lamp were coupled through small holes with depth. That is, inside with flame and outside with flammable gas were contacted through only small holes. Davy published a paper on the research results in 1816 (Davy, 1816).

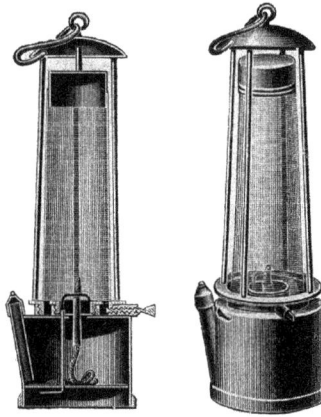

Fig. 4.7. Davy lamp.

4.3.4 *Miner's safety lamp*

In summer of the year coming back London, Davy was requested to study on miner's safety lamp which could light without explosion in colliery with fire-damp such as methane gas, and Faraday assisted his study during two months from October middle (Explanation 4.2).

4.3.5 *Joint research*

From 1818 to 1822, Faraday joint researched to improve quality of steel with Stodart (James Stodart) manager of surgical instrument maker. The joint research was on producing steel alloy with excellent hardness by adding little platinum and nickel. Then the Royal Institution had the most excellent facilities. This is the reason why joint research was requested from outside company in the Royal Institution. Under taking research on alloy, Faraday built up his reputation as a chemist.

Joint researching with outside company, Faraday got extra income. Because of increasing joint researches, incidental income exceeded original income. If the situation like this continued, he could accumulate wealth. However, during joint research, person or enterprise offering fund had leadership of joint research. Then Faraday's independence was lost, and he was aware that he could not research on the basis of his own thought. He gradually rejected contract of joint research, and devoted himself to primary research in the Royal Institution. Though he might be badly off, he selected the genuine scientist's way.

4.4 Oersted's Discovery

4.4.1 *Phenomenon of electromagnetism*

In 1820, Oersted (Hans Oersted) in Denmark discovered phenomenon of electromagnetism. As Fig. 4.8, when electric current (i) was shed in linear

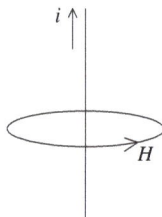

Fig. 4.8.   Relation between current (i) and magnetism (H).

Fig. 4.9.   Hans Christien Oersted (1777–1851) (portrait in 1832).

conductor, then magnetism (H) occurred with same direction as rotating direction of screw proceeding in direction of electric current. That is, the magnetic needle with direction of South–North changed its direction to direction perpendicular to wire.

4.4.2 *Reverse problem of Oersted*

Oersted's paper first was written in Latin, and due to its importance, was translated in all European languages. As Banks' successor Wollaston (William Hyde Wollaston) was inaugurated as temporary President until Davy's inauguration as President of the Royal Society. When Wollaston heard the discovery, he thought to grapple with "Reverse problem of Oersted's experiment." That is, he thought that if magnetic needle moved by wire live, then wire should move around magnetic needle, according to action-reaction law. He tried to realize his idea in the Royal Institution in the presence of Davy. Faraday seeing his experiment nearly was interested in the problem. Phillips who got acquainted with Faraday at lecture by Tatum mentioned above, recommended Faraday to write review on the paper by Oersted to "Annals of Philosophy." Therefore, Faraday repeated many experiments on the problem.

Skillful Faraday grappled with the problem by different method from Wollaston. As Fig. 4.11, using mercury (Hg) as liquid conductor, he set one end of wire to hinge (A), and made a movable wire (B), and dipped other end to mercury. He set a magnet (M) at center of container with mercury. When he contacted wires (B, C) to electric battery (V), wire (B) rotated around magnet (M). He called this phenomenon as "electromagnetic rotation." In Motor mentioned below (Paragraph 4.7 in this chapter), the coil rotates when electric current is shed to coil in magnetic field. In Faraday's electromagnetic rotation, wire rotates when electric current is shed to wire in magnetic field, and therefore Faraday's electromagnetic rotation was regarded as prototype of motor.

However, the trouble concerning originality on Faraday's success occurred. It was rumored that Faraday stole the part of Wollaston's research (James, 2010:39). Faraday published paper without quoting Wollaston's experiment. Therefore, it should seem for many persons that Faraday imitated Wollaston's experiment without regard to Wollaston. In fact, the reason for Faraday's action was that Faraday could not connect

Fig. 4.10.   William Hyde Wollaston (1766–1828) (portrait in 1820–1824).

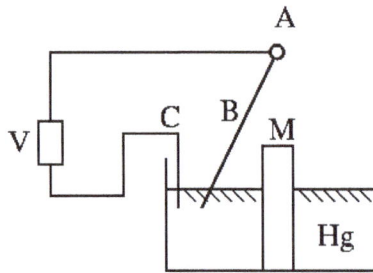

Fig. 4.11.   Faraday's method in reverse problem of Oersted. A: hinge, B: movable wire, V: electric battery, M: magnet.

with Wollaston, and he did not desire to quote Wollaston's research without permission (Bowers, 1978:44). Though Wollaston declined to pursue this trouble, the blame was continued afterward.

The bad feeling of controversy reached the zenith when Faraday was nominated as Fellow of the Royal Society. It was reported that Davy thought for Wollaston to be the first proposer of "Reverse problem of Oersted," and had an opinion slighting Faraday's originality on electro-magnetic rotation (James, 2010:39).

4.4.3 *Marriage*

On 20th February 1791, Faraday's father made his Confession of Faith in Sandemanian church, non Anglican church. On 15th July 1821, Faraday made his Confession of Faith in Sandemanian church as his father. On 12th June one month before the Confession, Faraday married Sarah (Sarah Barnard) who was Sandemanian and a daughter of silversmith. She was 9 years younger than Faraday and was modest. They lived at attic of the Royal Institution where Davy in his youth had lived, and afterward lived there during 37 years.

Faraday was kind knightly to her. She felt worth living by taking care of Faraday. Despite plain living, they lived a life filled with affection.

Fig. 4.12. Faraday' husband and woman.

4.5 Liquefaction of Chlorine Gas

In 1823, Faraday liquefied a gas (chlorine) for the first time. As Fig. 4.13, by Bunsen burner he heated chloric compound contained at end of sealed reverse V type of glass tube. Then sealed glass tube was filled with yellow chlorine gas produced by decomposition, and high pressure occurred. Chlorine gas was liquefied at the other end with low temperature. The experiment indicated that "gas was the vapor of liquid with low boiling point." Then, Faraday was aware of that "high pressure and low temperature" played important role for liquefaction of gas (Historical significance of liquefaction of chlorine gas by Faraday is described in Appendix 4.1). The trouble occurred about originality concerning Faraday's success of liquefying chlorine. Davy thought that he gave suggestion about Faraday's experiment, and he should receive the honor of discovery of liquefaction of gas. The feeling like this of Davy reached the zenith as bad feeling of controversy on originality of "Reverse problem of Oersted" when Faraday was nominated as Fellow of the Royal Society.

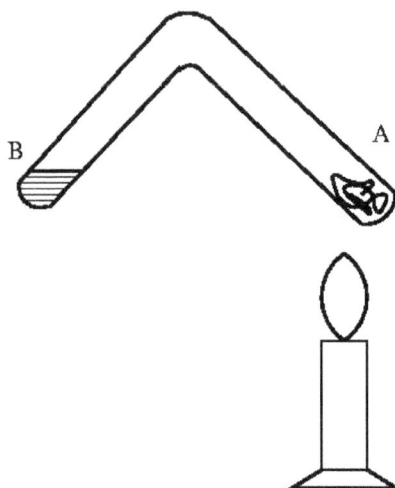

Fig. 4.13.   Liquefaction of chlorine gas. A: chloric compound, B: liquid chlorine.

Explanation 4.3 Liquefaction of gas

Phase diagram

As Fig. 4.14 phase diagram expresses the state of matter with pressure P and absolute temperature T. Vertical axis represents pressure P, and horizontal axis represents absolute temperature T. Absolute temperature's unit is Kelvin (K), and is expressed with centigrade (C) as follows:

Absolute temperature (K) = 273.15 + Centigrade (C).

Because absolute temperature is greater than 0 K, we can think only temperature range above 0 K using absolute temperature.

The solid line in Fig. 4.14 represents boundary between states of matters. The solid line above triple point E represents boundary between solid and liquid. There solid equilibrates with liquid. From this, it is known that there is a melting point where solid melts. The solid line below triple point E represents boundary between solid and gas, and there solid equilibrates with gas. Sublimation where solid becomes gas, occurs.

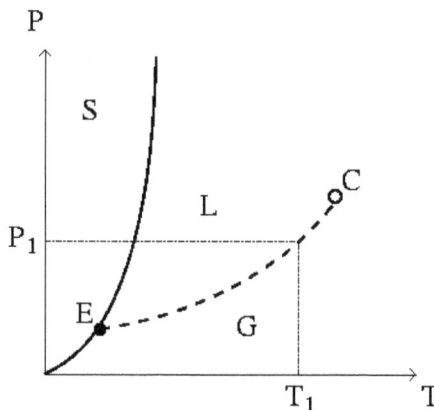

Fig. 4.14.   Phase diagram. S: solid, L: liquid, G: gas, C: critical point, E: triple point, P: pressure, T: absolute temperature.

The broken line in Fig. 4.14 represents boundary between liquid and gas, and there liquid equilibrates with gas. It is known that there is boiling point. The upper end C of broken line represents critical point, and above this point, liquid surface which distinguishes liquid from gas, is extinguished. When pressure is increased at temperature T_1 below critical point as Fig. 4.14, gas is liquefied. At the temperature above critical point, increasing pressure does not induce liquefaction of gas. The broken line is called vapor pressure curve.

Multistage cooling

The multistage cooling is described in the following. At T_1 we liquefy matter A with critical point above T_1 by increasing pressure as Fig. 4.15. Gas in vessel of liquefied matter A is evacuated by vacuum pump. By decreasing pressure by evacuating, the temperature of matter A falls according to vapor pressure curve. On decreasing temperature till T_2, at T_2 we liquefy matter B with critical point above T_2 by compressing. By decreasing pressure with evacuating gas in vessel of liquefied matter B, temperature of matter B can be cooled till T_3. The cooling like this is

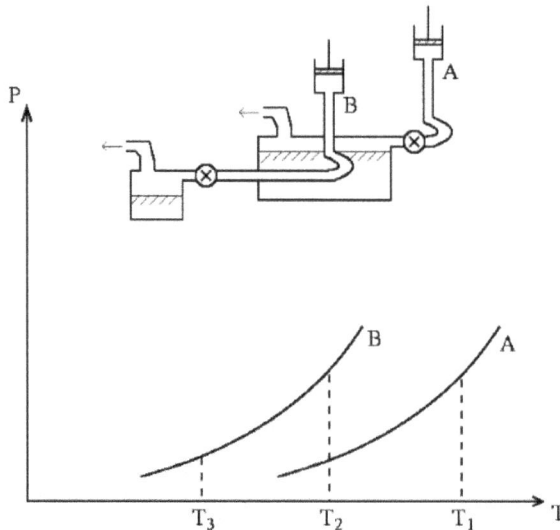

Fig. 4.15.   Multistage cooling. P: pressure, T: absolute temperature.

called multistage cooling. Onnes (Heike Kamerlingh Onnes) liquefied air using multistage cooling with Methyl Chloride, Ethylene, Oxygen and air.

Joule-Thomson effect

In 1908, preparing liquefied air and liquefied hydrogen, Onnes succeeded liquefaction of Helium gas using Joule-Thomson effect. The procedure is as follows: First he liquefied hydrogen gas cooled in advance by liquefied air, next using Joule-Thomson effect, he liquefied helium gas cooled in advance by liquefied hydrogen. Joule-Thomson effect is the phenomenon that if gas is streamed through the tube containing padding with plenty of holes and is expanded, temperature of gas changes. Near liquefaction always cooling effect is remarkable in Joule-Thomson effect.

Appendix 4.1 Historical significance of liquefying chlorine gas by Faraday

High pressure and low temperature

Faraday's knowledge about "high pressure and low temperature" was utilized in "difficult liquefaction of gas" which was the important problem for realizing extreme low temperature half century after Faraday's liquefaction. For example, at Craco in Poland, when Wroblewski (Zygmunt Wroblewski) and Olszewski (Karol Olszewski) succeeded in liquefying Oxygen in 1883, Oxygen gas was liquefied by introducing Oxygen gas with high pressure to tube cooled by being soaked in liquefied boiling ethylene evacuated. In 1908, when Onnes in Leiden University in Netherlands succeeded in liquefaction of helium gas which was most difficult to liquefy, he utilized the multistage cooling of the method of Wroblewski and Olszewski. Thus, the knowledge of "high pressure and low temperature" discovered by Faraday played important role in liquefaction of gas.

Superconductivity

Onnes was not satisfied with success on liquefaction of helium gas, and under the condition of low temperature realized by liquefied Helium gas, he tried to research property of matter. At experiment under extreme low temperature with liquefied helium he discovered "super conductivity" that resistance of mercury became zero under transient temperature. Superconductivity was an important discovery in the 20th century.

Because resistance of matter in superconductive state is zero, thermal loss is zero on shedding current and so great current can be shed. Then electromagnet with coil of superconductive matter can produce great magnetic field, and at present, the electromagnet using superconductivity is utilized in Linear Super express (Linear Motor Car) and medical apparatus magnetic resonance imaging (MRI) (Shioyama, 2002).

Cryogenics

In history of technology of "liquefying gas" which was important method realizing extreme low temperature, Faraday's "liquefaction of chlorine" was the pioneer. Dewar who succeeded in liquefaction of hydrogen gas, always respected Faraday who accomplished liquefaction of gas in the same Institution (Mendelssohn, 1971). The cryogenics utilizing cryogenic technology plays important role in modern society. History of cryogenics goes back to Faraday's works.

Fig. 4.16.   Heike Kamerlingh Onnes (1853–1926).

Fig. 4.17. James Dewar (1842–1923).

Explanation 4.4 Superconductivity

BCS theory

In 1911, Onnes in the University of Leiden discovered the superconductivity (zero electric resistance) at several kelvin (extreme lower temperature). In 1933, Meissner and Ochsenfeld (Meisner, 1933) discovered the complete diamagnetism. In 1957, Bardeen, Cooper and Schrieffer (Bardeen *et al.*, 1957) published the theory of superconductivity which contributed to the understanding the superconductivity at the extreme low temperature.

Cooper pair and attractive force between electron

It seems that two electrons repell and do not attract each other. However in 1956, Cooper (Cooper, 1956) indicated that a pair of electrons having equal and inverse momentum attract each other.

As Fig. 4.18 an electron attracts lattices with positive charge, creating the region of higher density of positive ions. As a result, other electrons are attracted to this region with higher density of positive ions, yielding

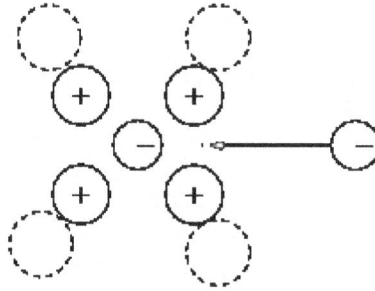

Fig. 4.18.   The region of higher density of positive ions.

the attraction two electrons (a pair of electrons). This pair of electron is called Cooper pair.

Meissner effect

On application of external magnetic field to a superconductor, a diamagnetic current is induced in the material, preventing from magnetic field intruding into the material. This phenomenon is called Meissner effect by which the magnetic field cannot intrude to the material.

Josephson effect

In 1962, Josephson predicted the superconductive current $\mathbf{J}_s$ due to tunnelling of Cooper pair at tunnel junction (the current $\mathbf{J}_s$ is called Josephson current), and the alternating current when voltage applied at the tunnel junction. The Josephson current is given by

$$\mathbf{J}_s = (eh/m^*)|\Psi|^2\{\nabla\phi - (2\pi/\Phi_0)\mathbf{A}\} \tag{1}$$

where h is Planck constant, $\mathbf{A}$ is the vector potential, $|\Psi|^2$ is the density of Cooper pair, ϕ is the phase of Cooper pair whose wave function is given by

$$\Psi = |\Psi|\exp\{j\phi\}, \qquad j = (-1)^{1/2} \tag{2}$$

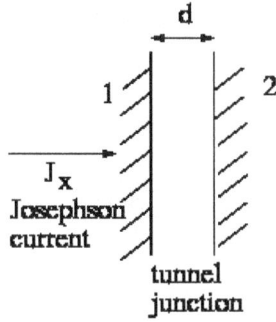

Fig. 4.19. Tunnel junction.

$m*$ is the mass of Cooper pair which is given by $m* = 2m$, m: mass of electron, c is the velocity of light. Φ_0 is magnetic flux quantum. ∇(nabra) is the spatial gradient vector whose component is $(\partial/\partial x, \partial/\partial y, \partial/\partial z)$. We consider the Josephson current J_x in case of Fig. 4.19.

J_x is given by

$$J_x = (eh/m*)|\Psi|^2 \{(d\phi/dx) - (2\pi/\Phi_0)A_x\} \tag{2}$$

Assuming that J_x and $|\Psi|$ are constant within the range of tunnel junction, and integrating (2) in the range within the junction, we have

$$J_x d = (eh/m*)|\Psi|^2 \{(\phi_2 - \phi_1) - (2\pi/\Phi_0)\int A_x \, dx\} \tag{3}$$

Define as

$$J_0 = (eh/m*d)|\Psi|^2 \tag{4}$$

$$\gamma = (\phi_2 - \phi_1) - (2\pi/\Phi_0)\int A_x \, dx \tag{5}$$

Limiting $d \to 0$, the second term in right-hand side of (5) is zero. Then we have

$$J_x = J_0 \, \gamma, \qquad \gamma = (\phi_2 - \phi_1) \tag{6}$$

The difference of phase $\gamma = (\phi_2 - \phi_1)$ is defined as γ being equel to $\gamma + 2\pi n$ (n: integer), therefore we have

$$J_x = J_0 \sin\gamma \qquad (7)$$

4.6 Election for Fellow of the Royal Society

4.6.1 *Nomination by Phillips*

In 1823, Faraday was nominated for Fellow of the Royal Society. To be Fellow of the Royal Society was honorable for scientist. In 1822, chemist Phillips who was a friend of Faraday since they attended lecture by Tatum, was already elected for the Fellow. Phillips nominated Faraday.

Though there was no rule, in accordance with the precedents, beforehand the proposer usually consulted with the president about nomination. Then already Davy became the President of the Royal Society. But without beforehand consultation, Phillips nominated Faraday. Davy got angry, and asked Faraday to refuse nomination. Faraday rejected saying that "only the proposer could refuse the nomination" (James, 2010:40).

Banks the preceding President, elected many persons who were not scientists by any means, utilizing his position. Since Davy was inaugurated as President, he endeavored to reform it. But for the strict reform concerning election of Fellow there was discontent in some party in the Royal Society. Consequently, Davy was anxious for opinion in the Royal Society. The official reason of Davy's opposition to electing Faraday was as follows: "Davy did not desire for him to be thought to agree nominating Faraday favoring his own disciple" (James, 2010:40). On 8th January 1824, Faraday was officially elected as Fellow of the Royal Society. There was only one opposition to Faraday's election. The opposition was Davy's opposition. Faraday was worried about Davy's intention. In February 1825, Faraday was promoted to the head of Laboratory in the Royal Institution. This promotion was proposed by Davy. Then, Faraday's worry about Davy's intention was eased. Then, Davy was away from the Royal Institution because of journey to the Continent and islands of Great Britain. During his absence, Davy made Faraday be agent in place of Davy.

Appendix 4.2 Benzene

In 1865, Kekule (Friedrich August Kekule von Stradonitz) proposed the structure of Benzene consisting of carbon and hydrogen (Kekule, 1865:98–110). As Fig. 4.20 the structure has hexagonal ring, and carbons are placed at the corner of hexagonal ring, and each carbon connects to neighboring carbon with single and double connection, and one hydrogen connects to all carbons. For example, styrene which is raw material of synthetic resin, and synthetic rubber, have Benzene ring. Benzene ring afterward would play the important role in organic chemistry.

4.6.2 *New substance*

In 1825, Faraday discovered a new substance. He discovered that the substance consisted of two elements, and found the ratio in the compound, and called the substance "Bi-carburet of hydrogen compound." Afterward, Mitscerlich (Eilhard Mitscerlich) called the substance "Benzene" (Appendix 4.2).

4.6.3 *Friday evening discourses and christmas lecture*

Since 1825, Faraday started to deliver Friday Evening Discourses. The persons who could attend the lecture were only the Royal Institution

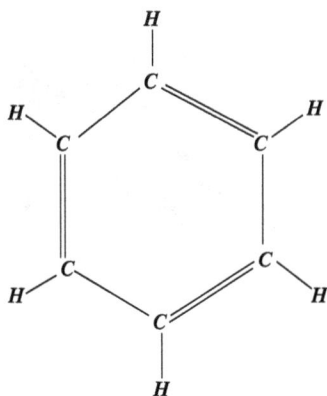

Fig. 4.20. Structure of Benzene.

Members and their guests. Next year, he started to deliver the Christmas Lecture. In 1861, his final Christmas Lecture was published as *The Chemical History of a Candle*. This book was the most popular among books published till then. In 1854, when Faraday delivered a lecture on education, husband of Queen Victoria, Prince Albert attended. Prince Albert who was interested in application of science, afterward, attended Faraday's lecture frequently, and in 1855 he attended Christmas Lecture with his two sons.

Faraday delivered 127 Friday Evening Discourses, and 19 Christmas Lectures. As writer Eliot (George Eliot) commented, Faraday's lecture explained state of the art science entertainingly as opera. Citizens listening his lecture were attracted to his personality, and had friendly feeling toward him. These two lectures are continued in the Royal Institution at present.

Fig. 4.21. Faraday delivering the Christmas Lecture (painting in 1855 by Alexander Blaikley).

4.7 Discovery of Electromagnetic Induction

4.7.1 *Transforming electricity to magnetism*

In 1825, Sturgeon (William Sturgeon) made an electromagnet by winding coil around a soft iron piece. Mankind could get a method of transforming electricity to magnetism.

In 1820s Faraday thought "Can we realize reverse transformation?" and wrote down idea of "transforming magnetism to electricity" in his notebook, and grappled with the great problem. Because then mankind not yet had the solution.

Though in 1827 he was invited as Professor of Chemistry in London University, he declined it, and decided to research the great problem in the Royal Institution. In May 1829, Davy passed away in Geneva. The friendly relation between Davy and Faraday was ended when Faraday was elected as Fellow of the Royal Society because of the trouble on originality in "the reverse problem of Oersted" and the originality on "liquefaction of chlorine gas."

4.7.2 *Electromagnetic induction*

In order to transform magnetism to electricity, during several years, he pocketed bar magnet and coil of wire, and always considered concerning

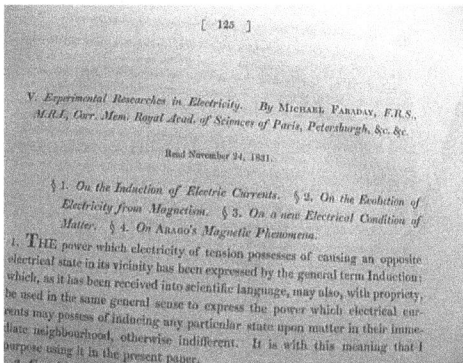

Fig. 4.22. Faraday's paper on electromagnetic induction (Faraday, 1832: 125–162).

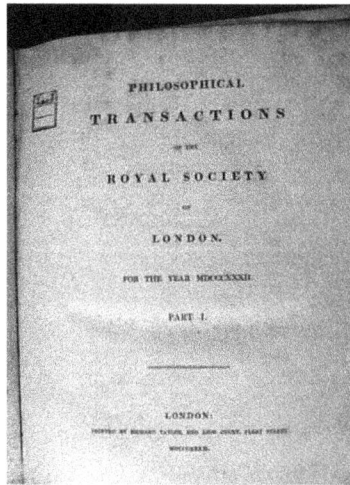

Fig. 4.23.   Academic Journal carring paper on electromagnetic induction (Philosophical Transaction of the Royal Society, submitted in 1831, Published in 1832).

electricity and magnetism. On 17th October 1831 he discovered electromagnetic induction that temporal change of magnetism induced electric current. The first paper by Faraday on the electromagnetic induction was submitted on 24th November 1831. Then mankind got a method for transforming magnetism to electricity which was inverse of electromagnet by Sturgeon. Electromagnetic induction had induced inventions of motor, generator and transformer that were the origin of modern convenience. In generator, temporal change of magnetism at ciol rotating in magnetic field by power of water or vapor induces electric current. On the other hand, in motor the rotation of live coil in magnetic field induces power. Motor is utilized to wide spread fields such as houshold electric appliances such as washing machine, communication instrument such as computer, vehicle such as automobile and industrial instrument such as lift.

Faraday researched on universal law on electromagnetic induction, and got conclusion that it is necessary for wire to cross magnetic line of force so as to occur temporal change of magnetism for purpose of inducing inductive voltage. Inductive voltage is proportional to the number of crossing magnetic line of force, whether wire crosses magnetic line of force perpendicularly or obliquely.

Explanation 4.5 Experiment with induction ring

As Fig. 4.24 Faraday wound the first coil at the one side of soft iron ring, and battery was connected to the coil, and current was switched on or off. At the other side of ring, he wound the second coil, and made closed circuit with wire connected to the second coil. Near the wire connected the second coil, magnetic needle was set. If current was induced then the magnetic needle should swing due to phenomenon discovered by Oersted. He tought that magnetism was induced in the soft iron by shedding current to the first coil, and the magnetism reached to soft iron penetrating the second coil, and it might induce current in the second coil. However, when he switched on shedding current, there occurred no change.

Many times, he switched on but during shedding current there occurred no change. When he switched off, thinking unsuccessfulness, he noticed that magnetic needle swang slightly. He did not miss occurrence in an instant. He paid attention to the instant of switching off. Certainly on switching off, magnetic needle swang. If scientific some fact is not verified under same condition and by any person and at any places, then the fact is not recognized as the truth. That is, reproducibility must be verified. Faraday verified the reproducibility.

At the process of switching, he was aware of important fact. On switching on, magnetic needle swang (James, 2010:57). He concluded that only when the stage of current at the first coil changed, at the second coil, current was induced.

The change of current at the first coil introduced change of magnetism in soft iron. The change of magnetism induced current at the second coil. Thus Faraday thought. Thus thinking, he estimated that magnetic change by other method should induce current.

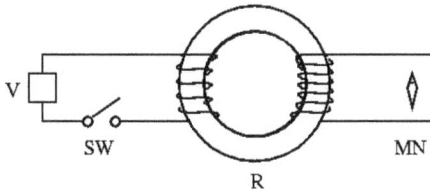

Fig. 4.24. Electromagnetic induction. R: induction ring, MN: magnetic needle, SW: switch, V: battery.

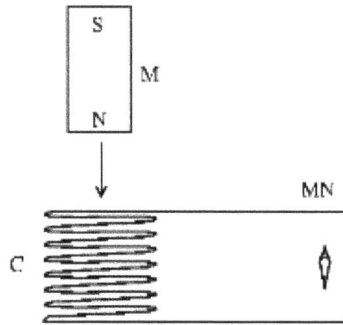

Fig. 4.25. Magnet's relative movement to coil. C: coil, M: magnet, MN: magnetic needle.

As Fig. 4.25, he wound linear coil and near the wire connecting to the coil for making close circuit, he set magnetic needle which played the role of detecting current. On the center axis of coil, he moved magnet near and far.

On 17th October 1831, magnetic needle swang as he predicted. He found that intense movement of magnet up and down, induced large swing of magnetic needle. That is, it was found that for the purpose of inducing electricity, it was necessary for magnet to relatively move, and temporal change of magnetism induced current.

The direction of induced current and change of magnetic field

The current is induced in direction for preventing the sum of magnetic flux density from changing.

Appendix 4.3 Magnetic line of force and electric line of force

Mathematical-physicists in the Continent did not admitted Faraday's concept on field of "line of force" such as magnetic line of force and electric line of force, because they believed electrodynamics of Ampere whose

electrodynamics took the form of remote action in mathematical formulation, but Faraday's concept was based on proximate action through
medium.

However, only Maxwell (Paragraph 5.6) admitted Faraday's concept.
Immediately after graduation of Cambridge University, he undertook
mathematicalization of concept of Faraday's "line of force." In 1855, he
submitted paper "On Faraday's lines of force" to the Cambridge
Philosophical Society, and it was published next year. Maxwell recognized Faraday's research as "the nucleus of everything electric since
1830" (James, 2010:89).

Furthermore, Einstein said "the electric field theory of Faraday and
Maxwell represented probably the most profound theoretical founding
which had been experienced by the foundations of physics since Newton's
time" (James, 2010:90).

4.7.3 *Invention of magnetic generator*

Faraday invented "magnetic generator" as Fig. 4.26 which induced continuous direct current with radial direction from B to A on rotating copper
board. The reason for the induced current is as follows: when the copper
plate rotates in direction C as Fig. 4.26 and magnetic flux density is
applied in direction from front to backward, then the movement of the part
AB of closed circuit results in increasing the sum of magnetic flux density
within the circuit, hence the current with direction from B to A is induced

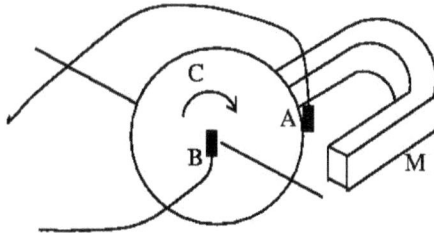

Fig. 4.26. Magnetic generator by Faraday. M: magnet, A, B: terminals sliding contacting, C: rotating copper board.

for preventing the sum of magnetic flux density from increasing. This was the prototype of dynamo (direct current generator with commutator).

In alternating current generator, direction of current is inverted every half rotation of coil in magnetic field. To always get direct current with constant direction, in dynamo, two fixed terminals which are sliding contacting to rotating commutator and connecting outer circuit, are set. Every half rotation, by changing connection to coil, direct current with constant direction is taken out to outer circuit. Commutator consists of two metal boards each of which is put on rotating shaft for two boards to go around rotating shaft (Explanation 4.6).

Explanation 4.6 Dynamo

As Fig. 4.27, in dynamo, at the right hand side of a coil rotating in anti-clockwise in the magnetic field **B**, the current i flows in direction shown as Fig. 4.27, because on coil rotating, the sum of B within coil is increased, hence the current **i** is induced for preventing the sum of B from increasing. Therefore a fixed terminal S1 sliding contacting to rotating commutator R2 is a negative (–) terminal. After half-rotation, commutator R1 will sliding contact to a fixed terminal S1. At a coil connecting R1, current i flows because the coil is at the right hand side, then S1 becomes negative (–) terminal. Thus, a fixed terminal S1 is always negative (–) terminal, and S2 becomes positive (+) terminal, then a direct current is obtained.

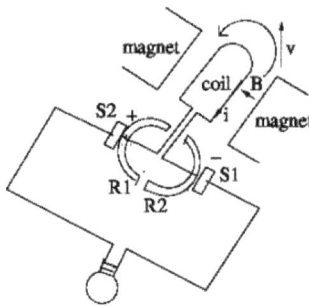

Fig. 4.27.   Dynamo; B: magnetic flux density, v: velocity of coil, i: current, S1, S2: fixed terminals, R1, R2: commutators (connecting coil).

Explanation 4.7 Theorization of electromagnetic phenomena by Maxwell

In 1864, Maxwell derived electromagnetic equations which were expressed as four partial differential equations unifying all electromagnetic phenomena. One equation among them corresponds to electromagnetic induction discovered by Faraday, and it consists of the first term (left hand side of equation) expressing rot**E**: partial derivative with respect to spatial coordinate of electric field **E** and the second term (right-hand side) expressing– ∂**B**/∂t: the negative value of temporal change of magnetic flux density **B**. **B** (= μ**H**, **H**: magnetic field, μ: magnetic permeability) is called also magnetic induction. Assume that coil of one turn is formed with wire, and upward **B** temporally increases.

When each term is integrated for area within coil, the first term becomes line integral along coil of electric field, and expresses inductive voltage. The reason is as follows:

We integrate **E** along peripheral lines of rectangle S as Fig. 4.28, where a coil is assumed to be in *x–y* plane.

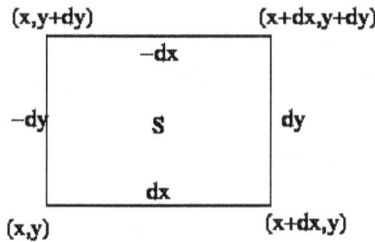

Fig. 4.28. Periphery of S with area *dxdy*.

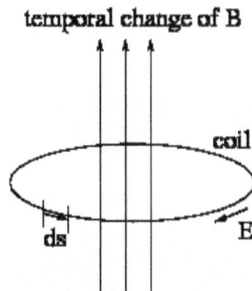

Fig. 4.29. **B** through the area of coil.

Then we have a line integral of **E** along periphery of S:

$$\int (\mathbf{E} \cdot \mathbf{ds}) = E_x(x,y,z)dx + E_y(x + dx,y,z)dy - E_x(x,y + dy,z)dx$$

$$-E_y(x,y,z)dy$$

$$= [E_x(x,y,z) - E_x(x,y + dy,z))]dx$$

$$+ [E_y(x + dx,y,z) - E_y(x,y,z)]dy$$

$$= [\partial E_y/\partial x - \partial E_x/\partial y]dxdy \ (\text{when } dx,dy \to 0)$$

$$= (\text{rot } \mathbf{E})_z \, dxdy$$

where E_x is the *x*-component of **E**, **ds** is the segment of peripheral line, and (**E** · *ds*) is the scalar product of **E** and **ds** which is $E_x dx$ for **ds** = *dx*, and $E_y dy$ for **ds** = *dy*. (rot **E**)$_z$ defined as $[\partial E_y/\partial x - \partial E_x/\partial y]$ is the *z*-component of vector rot **E**. The partial derivative $\partial E_y/\partial x$ is defined as lim $[E_y(x + dx, y, z) - E_y(x, y, z)]/dx$ (when $dx \to 0$). The above equation indicates that the area integral of (rot **E**)$_z$ within the area S with area: *dxdy* is the line integral of **E** along the peripheral line of area S. As Fig. 4.30, the line integral along line Σ is the sum of line integrals along lines Σ_i, $i = 1, 2$, which constitute Σ. Hence the integral of (rot **E**)$_z$ over the area of coil equals to integral of (**E** · *ds*) along coil (Stokes' theorem), giving the inductive voltage V because *dV/ds* = E, ds: the line segment in integral as Fig. 4.29.

The second term expresses the temporal change of sum of magnetic flux density **B** penetrating coil. That is, the above equation is the formulation of electromagnetic induction discovered by Faraday. Other three equations are explained in Explanation 5.2.

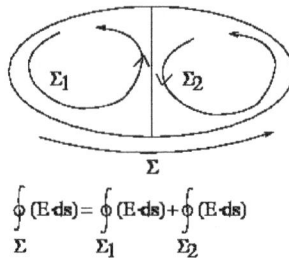

$$\oint_{\Sigma} (\mathbf{E} \cdot \mathbf{ds}) = \oint_{\Sigma_1} (\mathbf{E} \cdot \mathbf{ds}) + \oint_{\Sigma_2} (\mathbf{E} \cdot \mathbf{ds})$$

Fig. 4.30.   Line integrals.

Fig. 4.31.   James Clerk Maxwell (1831–1879).

The curves with directions of electrical displacement **D** (= ε**E**, ε: dielectric constant (Explanation 4.10)) and magnetic flux density **B** are, called as electric line of force and magnetic line of force, respectively.

Electromagnetic induction discovered by Faraday is a phenomenon where if the magnetism changes at place with electric circuit such as coil, that is, if the state of magnetic line of force changes, power is induced. We call the power as inductive voltage, and call the current as inductive current. As Appendix 4.4, the law of electromagnetic induction is derived on the basis of the behavior of electron in metal moving in magnetic field.

Appendix 4.4 Modern interpretation of electromagnetic induction

An electron with charge $-e$ moving at velocity **v** in magnetic flux density **B** receives a force **F** called Lorentz force as Fig. 4.32. The direction of Lorentz force is perpendicular to the plane containing **v** and **B**, and

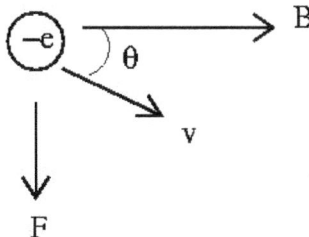

Fig. 4.32.   Lorentz force.

magnitude of Lorentz force is $evB \sin\theta$ with sinusoidal function of angle θ between **v** and **B**. Where v is the magnitude of velocity and B is the magnitude of magnetic flux density.

In metal, there are a free electron which can move free and has negative charge, and positive ion with positive charge which cannot move freely (a positive ion is an atom in crystal removed a free electron). When wire moves at velocity **v** in field of magnetic flux density **B** as Fig. 4.33, an electron in wire receives Lorentz force. Because the free electron moves to the end Q of wire, at P of wire positive charges accumulate and at Q of wire, negative charges accumulate, and the electric field **E** (force influencing unit charge, that is, negative spatial gradient of potential) for electron with direction from Q to P, occurs.

Because the electric field prevents electron from streaming without limit to end Q by Lorentz force, there is the state of equilibrium keeping balance of Lorentz force and electric field. That is, the force $e\mathbf{E}$ due to electric field for electron and Lorentz force $ev\mathbf{B} \sin\theta$ balance. Because magnitude of electric field is spatial gradient of voltage, magnitude of electric field is given by voltage between P and Q, divided by length d between P and Q. Because voltage is given by the product of length d and magnitude of electric field, voltage is given by $dv\mathrm{B} \sin\theta$.

If wire is connected to resistor R or wire, then current i flows as Fig. 4.34. Assume that wire PQ in Fig. 4.35 perpendicular to parallel sides of U-shape wire moves at velocity **v**, and velocity **v** is perpendicular to **B**

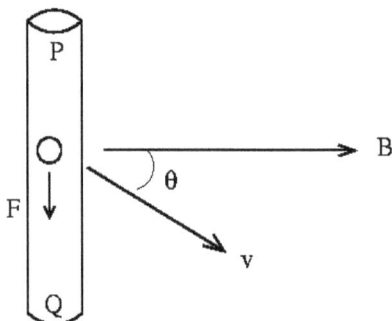

Fig. 4.33. Inductive voltage.

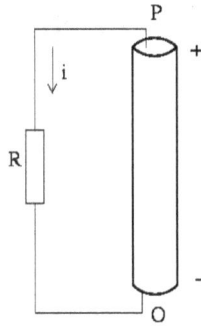

Fig. 4.34. Current due to inductive voltage.

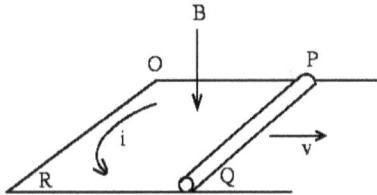

Fig. 4.35. Sliding movement of wire PQ. **B**: magnetic flux density, *v*: velocity, **i**: inductive current.

and *PQ*'s direction. The length of *PQ* is *d*. Then at *PQ* power *dv*B is induced, the current is shed with direction of *P*→*O*→*R*→*Q*.

On the other hand, during the time *t*, the area inside of *PORQ* changes with *dvt*, and change of magnetic flux (the sum of magnetic flux density in PORQ) is given by *dvt*B. Therefore, considering inductive voltage to be *dv*B, the following relation is obtained.

Inductive voltage = change of magnetic flux per unit time

This relation is law of electromagnetic induction. Magnetism due to inductive current has inverse direction to **B**. Thus, it is found that in magnetic flux density **B**, the current is induced so as to yield magnetism with opposite direction to **B**.

In electromagnetic induction, the current is induced in direction for preventing the sum of B within electric circuit from changing.

4.8 Discovery of Laws of Electro-chemical Decomposition

Voltaic electric battery was a pile of the first conductors such as copper, inserted by the second conductor such as electrolyte. Hence the battery was called Voltaic pile (Appendix 4.5). In 1800, Volta informed this invention to Banks President of the Royal Society.

Banks who received a letter from Volta, informed the invention of Voltaic pile to Nicholson (William Nicholson) English chemist and Carlisle (Anthony Carlisle) Surgeon. Two persons fabricated Voltaic pile and utilized the battery as electric power. As Fig. 4.36, two wires which were electrodes connected to terminals of battery, were dipped in the other vessel filled with water. Then, hydrogen occurred at cathode wire and oxygen occurred at anode wire, that is, electro-chemical decomposition was verified. Nicholson and Carlisle were the person who verified the chemical reaction for the first time.

In 1811, Faraday verified the idea proposed by Gay-Lussac (Joseph Louis Gay-Lussac) and Thenard (Louis Jaques Thenard) "The factor controlling the rate of decomposing electrolyte is not concentration of solution and electrode, but intensity of current through solution." That is, it was concluded that in electro-chemical decomposition, the connection of electrode with solution was not necessary, and process of electro-chemical decomposition did not depend on the action of electrode, but depend on the intensity of current passing through solution.

4.8.1 *Laws of electro-chemical decomposition discovered by Faraday*

In 1833, Faraday discovered the first law of electro-chemical decomposition "the quantity of product due to electro-chemical decomposition is

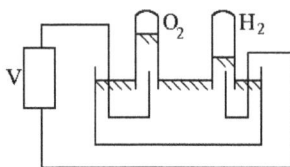

Fig. 4.36. Electro-chemical decomposition of water. V: battery, O_2: oxygen gas, H_2: hydrogen gas.

proportional to the absolute quantity of electricity passing through solution" (Explanation 4.8). Furthermore, measuring mass of constituent obtained by decomposition, he found that quantities of constituents produced at cathode and anode indicated constant ratio. The quantity determining this ratio was called "electric chemical equivalent." From measuring the chemical equivalent, Faraday discovered the second law of electro-chemical decomposition "electric chemical equivalent coincides chemical equivalent and the ratio of quantities of products by decomposition is equal to the ratio of chemical equivalent" (Explanation 4.8).

Appendix 4.5 Invention of Voltaic pile

In 1780, Galvani (Luigi Galvani) Professor of Anatomy in Bologna University dissected a frog, and contacted surgical knife to nerve of frog. Then, suddenly muscle of leg intensively cramped (Yukawa & Tamura, 1955–1962:II). Galvani published his theory "Muscle of frog produces electricity." This discovery induced great influence to physicist and physiologist.

Volta repeated experiment of leg of frog by Galvani. He set silver foils at two separate points on nerve of leg of frog, and shed electricity from silver foils. Then, muscle of leg of frog cramped. From experiment, he

Fig. 4.37. Luigi Galvani (1737–1798).

Fig. 4.38.   Alessandro Volta (1745–1827) (portrait in the beginning of the 19th century).

recognized that "stream of electricity stimulated nerve, and as the second-ary effect from nerve, muscle cramped."

Furthermore, he accomplished other experiment with different method. He connected wire to two silver foils placed at two separate points on nerve and did not shed electricity. Then, muscle cramped inten-sively. Connecting wire to silver foils as electrodes, electricity due to forming closed circuit of nerve and wire, stimulated nerve, and as the secondary effect from nerve, muscle cramped. From this fact, he denied Galvani's theory.

Volta discovered that using two different metals instead of silver foils set at two points on nerve, more intensively muscle cramped. He reached an idea "the presence of two different metals is the cause of strong electric stream. Dissection of frog produces no electricity."

During proceeding experiment of electric stream using different met-als, Volta discovered that "contact of the first conductor such as zinc, cop-per and silver, and the second conductor such as wet conductor or fluid, causes electric stream." He discovered that the second conductor such as matter dipped in salt water (electrolyte) being inserted in two electrodes of the first conductors such as zinc and copper, causes electric stream. Piling up the combination where the second conductor was inserted in two the first conductors, Voltaic pile was invented.

Utilizing voltaic pile as direct current power supplier, Davy performed electro-chemical decomposition. In modern society, there are following

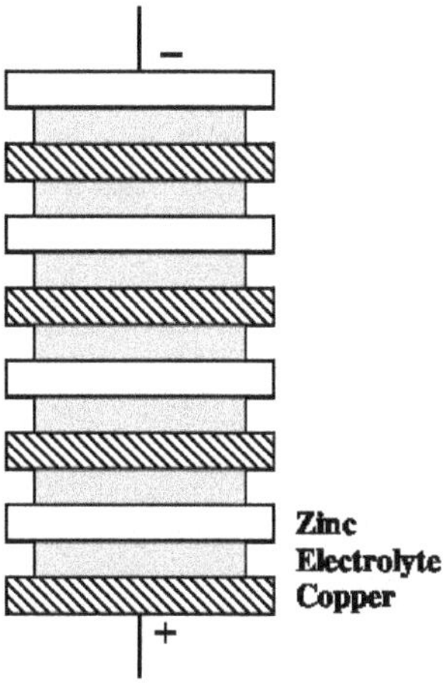

Fig. 4.39. Voltaic pile.

batteries. There is rechargeable battery such as lithium ion battery capable of charge and discharge. As other battery than that based on chemical reaction, there is a battery based on physical phenomenon such as solar battery which produces power on radiating semiconductor (p–n junction) (Shioyama, 2002).

According to the law, for example, the ratio of hydrogen and oxygen produced by decomposition of water, is the ratio of chemical equivalent 1.008:8.000, and the ratio of mol is 1:1/2, and the ratio of volume is 2:1 (Explanation 4.8).

Utilizing precipitation of metal in solution by decomposition, Davy was successful in isolating sodium and potassium for the first time. In modern society, using decomposition plating is performed.

In 1834, Faraday continued research on electro-chemical decomposition, and introduced new academic terms in order to express precisely the phenomena of electro-chemical decomposition and also to be easy to discuss among scientists, with consulting linguists. For example, new academic terms were electrode, anode, cathode, electrolyte, anion, cation and ion. These terms are used at present. On using such academic terms, metal in solution is expressed by anion.

Explanation 4.8 Laws of electro-chemical decomposition

Faraday's laws of electro-chemical decomposition are explained quantitatively.

The first law: quantity of constituent produced at electrodes by electro-chemical decomposition is proportional to quantity of electricity (ampere × second)

The second law: quantity of constituent produced by same electricity is proportional to chemical equivalent.
(**example**) The ratio of quantities of H_2 and O_2 produced by decomposition of water is the ratio of chemical equivalent 1.008:8.000. The ratio of mol is 1:1/2. According to Avogadro's law, the ratio of volume is 2:1, that is, volume of H_2 is double O_2.

The third law: quantity of electricity for precipitation of 1 gram chemical equivalent = 96500 coulomb = 1 faraday.
   1 coulomb = quantity of electricity given by 1 ampere × 1 second.
   1 gram chemical equivalent = quantity of constituent which is expressed by gram with numerical value of chemical equivalent.

Avogadro's law: there are same number of molecules under same temperature, same pressure and same volume.

Explanation 4.9 Chemical equivalent

The ratio of quantities of constituents produced at anode and cathode by electro-chemical decomposition is determined by the ratio of chemical equivalent. Chemical equivalent is explained in the following.
   Atomic weight represents relative mass, and the mass of one atom of oxygen O is 16.000 as basis. Constituent's quantity of atomic weight with gram is called gram atom. One gram atom of oxygen is 16.000 g. Quantity of molecule represents relative mass of molecule. Mass of one molecule of oxygen is 32.000 as basis. Gram molecule (mol) is the quantity of constituent or compound of quantity of molecule with gram. For example, gram molecule of nitrogen (N) of 1 mol is 28.016 g.

In one gram molecule (mol), there are Avogadro number ($6.02214076 \times 10^{23}$) of molecules.

Chemical equivalent is defined as constituent's quantity compounding oxygen 1/2 atomic weight (or hydrogen 1 atomic weight). For example, chemical equivalent of oxygen is 8.000 (from H_2O), and chemical equivalent of chlorine is 35.457 (from HCl).

When atomic weight and valence are known, chemical equivalent is given by (atomic weight)/(valence). Where valence is defined as number of hydrogen atom combining with one atom of element. For example, the valence of Cl is 1 (HCl), 2 for O (H_2O) and 3 for N (NH_3). The valence of constituent not combining hydrogen is known by the ratio of number combining other element with known valence.

4.9 Research on Dielectrics, Light and Magnetism, and Magnetic Substance

4.9.1 *Research on dielectrics*

On researching on electro-chemical decomposition, Faraday paid attention to the effect based on law of electric conduction. Electrolyte kept conductive force at liquid, but it lost the force at solidity. For example, when water became ice at solid state it could not stream current. When platinum foils were set at both upper and lower surfaces of ice, and connected foils to power, then electric charges were induced. If ice was liquefied to water, it streams current. He paid attention to the phenomenon like this.

The phenomenon where charges were induced on the surfaces of insulator set between two metals (electrodes) connected to power terminals, occurred by the action of polarized contiguous particles (dipoles) in all insulator medium. He thought that electric action between separate points, occurred through medium matter. Thus, Faraday related charge induction on surfaces of insulator medium, to polarization of insulator medium placed between electrodes (Explanation 4.10). In order to research the effect of polarization, he performed the experiment to seek how the charge induced on insulator medium between two metals (electrodes) depended on the insulator medium mediating between two metals.

Consequently, it was found that for the same voltage, the charges induced on surfaces of insulator medium were different for different insulator medium. Faraday called the relation between charge and voltage "dielectric capacity." This is related to "dielectric constant." In November 1837, he proposed to call insulator medium as "dielectrics."

Explanation 4.10 Dielectric constant

When as Fig. 4.40 two particles with electric charge q_1, q_2 are placed separately with distance r, the static electric force **F** influencing particles is proportional to the product of charge q_1 and q_2, and to the inverse square of distance **r**. If charges are same signs, then static electric force is repulsive force, and if charges are different signs, static electric force is attractive force. This law is called Coulomb's law (Yukawa & Tamura, 1955–1962:II).

Next, we consider molecules in dielectrics (insulator). If in a molecule, center of gravity of positive charge does not coincide to that of negative charge and separated mutually with distance, the molecule is called polar molecule. On the other hand, if both centers of gravity coincide, then molecule is called non-polar molecule.

When as Fig. 4.41 polar molecule is placed between upper and lower metal with negative and positive charge respectively, polar molecule rotates to the direction of static electricity due to metals with charges. On the other hand, when non-polar molecule is placed, between upper and lower metal, positive and negative center of gravity move to direction of static electricity, and molecule becomes dipole.

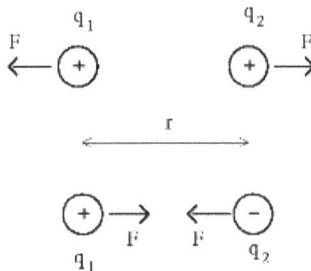

Fig. 4.40.   Coulomb's law.

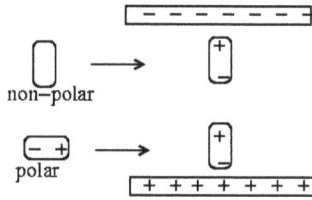

Fig. 4.41. Molecule in static electricity.

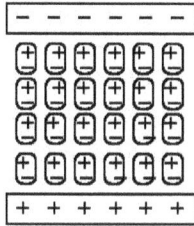

Fig. 4.42. Dielectric polarization.

That is, in the static electricity, both polar and non-polar molecules direct to static electricity as dipole. Consequently, from the direction of dipole in static electricity, on upper surface of dielectrics placed in static electricity, positive charge appears and on lower surface, negative charge appears as Fig. 4.42. Phenomenon like this is called "dielectric polarization."

When the voltage between upper and lower electrodes is V as Fig. 4.43, charge Q induced on surface of dielectrics placed between two electrodes, is proportional to V. The proportional coefficient C is called "Capacity."

$$Q = CV.$$

Denote by C_0 the capacity in case of vacuum, the capacity C of dielectrics is expressed by

$$C = \varepsilon C_0$$

The coefficient ε is called dielectric constant.

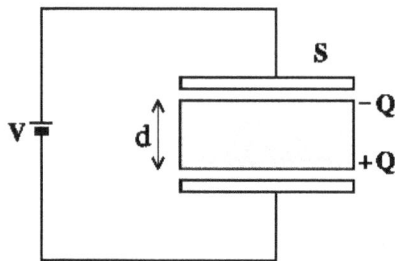

Fig. 4.43. Condenser. *V*: voltage of power supply, *S*: area of electrode, *d*: distance between electrodes, *Q*: induced charge on surface of dielectrics.

As Fig. 4.43, when the area of electrodes of condenser (called also capacitor) is *S*, distance between electrodes is *d*, capacity *C* is proportional to ratio *S/d*. This proportional coefficient is dielectric constant.

$$C = \varepsilon S/d, \qquad \varepsilon: \text{dielectric constant.}$$

Condenser is an important electronic part consisting electronic circuit together with resistor and coil. Using dielectrics with large dielectric constant, we can fabricate condenser with large capacity.

Condenser has a property that it does not pass through direct current, but pass through only alternating current. Therefore, in electronic circuit, condenser is utilized as high pass filter for the purpose of passing through alternating current of high frequency. High pass filter passes through alternating current with higher frequency than threshold. The threshold is called "cutoff frequency," and is inverse proportional to capacity of condenser. On the other hand, when direct current is treated in electronic circuit, smooth circuit is used in order to remove unnecessary ripple in signal. In the smooth circuit, condenser is used in order to remove ripple by falling it to earth. When the capacity of condenser in the smooth circuit is larger, the function of smooth is the better.

Particles in dielectrics can be compared to system of small magnetic needles. Faraday came to an idea that the polarization of dielectrics is similar to polarization in soft iron in magnetic field. That is, he thought that polarization of dielectrics could be explained by electric field. Faraday introduced a idea of electric line of force. Thus, the concept of "electric field" was born. The concept of electric field was utilized in theorizing electromagnetic phenomena by Maxwell (Explanation 4.7, Appendix 5.1).

As the unit of capacity of condenser (also called capacitor), farad (F) is used. This was named next year after he passed away for the purpose of praising his contribution.

In November 1839, Faraday was under poor condition, and suffered from dizziness and headache. In December 1840, the Royal Institution exempted him from duty till his recover. In 1841, he took a rest for three months in Swiss. But, he did not completely recover.

4.9.2 *Research on light and magnetism*

As Fig. 4.44, light is transverse wave vibrating in the direction perpendicular to the running direction. The direction of vibration is any direction in the plane perpendicular to the running direction. But, by passing through special crystal, light becomes a polarized light with constant direction of vibration.

In August 1845, Thomson (William Thomson) Faraday's friend mathematical physicist inquired to Faraday by a letter as "what effect a transparent dielectrics would have on polarized light" (James, 2010:79). For answering to the inquiry, Faraday tried experiment.

Believing that magnetic and electric action were correctly explained by the concept of line of force. Faraday tried to understand magnetic property by magnetic line of force. To examine relation between light and magnetism, he used heavy glass as polarizer, and set powerful electromagnet near the heavy glass, and light was passed through heavy glass. Then, the direction of polarization rotated. The rotation of polarization of light

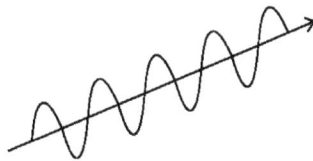

Fig. 4.44. Vibration of light.

Fig. 4.45. Electromagnet used in research on magneto-optical effect. (in Faraday Museum taken by Dr. Matsuda in September 2017).

due to magnetic field was called "magneto-optical effect: Faraday's effect." It was discovered on 13th September 1845.

As an application example of Faraday's effect, there is a method for measuring current in high voltage power-transmission line with non-contact safety. The current is measured utilizing change of rotation angle of polarization in Faraday's effect depending on the intensity of magnetic field induced by current passing through power-transmission line.

In 1862, Faraday performed experiment to research on change of wave length of light passing through strong magnetic field. However because the sensitivity of apparatus was insufficient, he was not successful. The phenomenon expected by Faraday was discovered by Zeeman (Pieter Zeeman) in 1896, and was called "Zeeman effect" (phenomenon where wave length of light through strong magnetic field, splits to multiple wave lengths). From the fact that the phenomenon expected by Faraday was discovered about 30 years later, it is found that he researched on basis of his acute intuition.

4.9.3 *Research on magnetic substance*

Faraday researched on the magnetic behavior of substance magnetized by magnetic field. He hung a substance in strong magnetic field so that the substance could freely move in space. In case of many substances such as glass, the substances moved to direction perpendicular to direction of magnetic line of force. But, in case of paramagnetism such as aluminum, the substance moved to direction of magnetic line of force. In many substances such as glass moved to direction perpendicular to magnetic line of force, because magnetization reverse to magnetic field occurred. Faraday called the magnetic property of substance different from paramagnetism as "diamagnetic" in 1845.

Explanation 4.11 Paramagnetism and diamagnetism

Paramagnetism is a form of magnetism whereby some materials are weakly attracted by an externally applied magnetic field, and form internal, induced magnetic field in the direction of the applied magnetic field.

In contrast with this behavior, diamagnetic materials are repelled by magnetic fields and form induced magnetic fields in the direction opposite to that the applied magnetic field.

Paramagnetism is due to the presence of magnetic dipole moment of molecule or atom which acts like tiny magnet, in the material. An external magnetic field causes the magnetic dipole moments to align parallel to the field, causing a net attraction. Paramagnetic materials include aluminium, oxygen, titanium, and iron oxide (FeO). In the absence of an externally applied magnetic field, paramagnetic material does not retain any magnetization because thermal motion randomizes the tiny magnet orientations.

Diamagnetism is the property of materials that are repelled by a magnetic field; an applied magnetic field creates an induced magnetic field in them in the reverse direction, causing a repulsive force. The magnetic permeability of diamagnetic materials is less than the permeability of vacuum. In most materials, diamagnetism is a weak effect which can be detected only by sensitive laboratory instruments, but a superconductor acts as a strong diamagnet because it entirely expels any magnetic field from its interior (the Meissner effect).

Diamagnetism was first discovered when Anton Brugmans observed in 1778 that bismuth was repelled by magnetic fields. In 1845, Michael Faraday demonstrated that it was a property of matter and concluded that every material responded (in either a diamagnetic or paramagnetic way) to an applied magnetic field. Faraday first referred to the phenomenon as diamagnetic (the prefix dia- meaning through or across), then later changed it to diamagnetism.

4.10 Social Contribution by Faraday

4.10.1 *Lighting*

As Faraday's social contribution, lighting relation is considered. In 1836, Faraday was inaugurated as adviser of Trinity House. He continued this responsibility till 1865. Trinity House was station responsible for safety of coast voyage, and changing old lighting was important work. Utilizing his ability, he planned the save of fuel consumption and improvement of efficiency of light house lighting.

In 1840, he invented a new chimney for oil lamp. By this chimney, gas occurring in oil combustion was efficiently removed and cloudiness of lamp glass decreased and lighting was improved.

The Faraday's new chimney for oil lamp was applied to not only English light house but also library and Buckingham Palace. The Times of

Fig. 4.46.   Faraday (portrait in 1842).

Fig. 4.47. Bust of Faraday by Mattew Noble, 1854. (in Faraday Museum, taken by Dr. Matsuda in September 2017).

English leading newspaper reported that the oil lamp lighted Princess Helena's baptism. Because he was not interested in patent, in 1842 he transfer the patent to his brother.

In 1854, Faraday was requested from Trinity House to test electric lighting system invented by Watson (William Watson) in 1852. The electric lighting system utilized arc discharge occurred between carbon electrodes supplied voltage by battery.

Faraday summarized test results in report with 4200 words, and indicated the following:

① The problem collecting chemical substances produced by battery should be solved.
② Wide room for battery was necessary.
③ The room for three person living for battery maintenance was necessary.
④ Brightness of arc discharge was variant in time.
⑤ It was difficult to keep persons maintaining this electric lighting system at the present stage.

Fig. 4.48.   William Watson (1715–1787) (print in1784).

He concluded that application of the electric lighting system to light house could not be recommended.

In 1857, Holmes (Frederick Holmes) proposed other electric lighting system using arc discharge. In case of Watson's system, battery was used, but in case of Holmes's system, generator driving by steam engine based on electromagnetic induction discovered by Faraday was used. By Faraday's positive recommendation, Trinity House approved budget of test run. On 8th December 1857, in the presence of Faraday, the electric lighting system lighted the English Channel for the first time. But afterward, because technical problem was found, Holmes's system was given up. The oil lamp was used till light house's lighting with incandescent light was invented in 1920 by Joseph Swan.

4.10.2 *Other social contribution*

From 1829 to 1851 he served Professor of Chemistry in English Military Academy.

In 1840, Faraday was elected as Elder of Sandemanian church. The works of Elder were preaching, baptizing infants, presiding at the

Love Feast. In 1860, he was again elected as Elder, and served the role till 1864. He was not eloquent. Due to his personality, the period of his Elder was continued long time. Before he was elected as Elder, he performed work supporting orphanage (in about 1832). For Faraday, church, family and work were related, and his living and work (science) were understood only by his religious belief and practice.

On 28th September 1844, in Haswell Colliery explosion occurred, causing 95 person's death including 3 boys at the age of 10s. He and Lyell (Charles Lyell) were requested to try the explosion accident from Prime Minister Peel (Robert Peel). On 8th October, they went to Haswell to investigate, and reported that it was important to improve the ventilation in colliery in order to prevent from being filled with fire-damp. In October, before they leaved Haswell to London they contributed to establishing a fund for supporting widow and orphan losing father.

In war of Britain-France against Empire of Russia (Crimean War), to prevent the Empire of Russia's from expanding to Western Europe due to declining of the Turkish Empire, Britain allying with France attacked Cronstadt on Baltic sea and Sebastopol on the Crimea peninsular, for the purpose of giving trade's damage. Faraday was asked often technical advice in secret from Royal Navy.

In 1854, Cochrane (Thomas Cochrane) proposed the use of Sulphur-filled fire ships on attacking Cronstadt. Faraday who was requested comment on the proposition, analyzed the poison gas chemically, and insisted that it was necessary to know the situation in case of using Sulphur-filled fire ships. He reported that chemical weapon should not be used. On the basis of Faraday's report, Graham (James Graham) the Minister of Navy dismissed Cochrane's proposition. Thus, considering always the real influence to citizen, Faraday gave cautious counsel.

4.11 A Grace and Favor House on Hampton Court Offered by Queen Victoria to Mr. and Mrs. Faraday

As mentioned above, since November 1839 Faraday suffered from poor condition. His wife also was under poor condition, and felt difficult to walk.

Fig. 4.49. Hampton Court Palace (photograph taken from https://www.britainexpress. com/attractions.htm? attraction=169).

Queen Victoria heard from Prince Albert that Mr. and Mrs. Faraday were living at attic in the Royal Institution for 37 years, and at that time, being under poor condition, they had difficulty of going up the stairs. She offered them to use in life a grace and favor house on Hampton Court facing the upper stream of the Thames. Since then, they lived at graceful residence in life.

Faraday was not interested in honor. Therefore, he declined President of the Royal Society twice. Also he declined President of the Royal Institution. He with no child was fear his wife's old age till his last moment. On 25th August 1867, he passed away quietly sitting at armchair, at 76 years of age. He was buried at Highgate Cemetery.

He was born as blacksmith's son. At 13 years of age, giving up regular schooling, Faraday started his life as a newspaper-cumerrand boy at bookshop and bookseller. He yearned to be scientific practitioner, and endeavored self-studying and got a position in the Royal Institution of Great Britain. Accomplishing plenty of epoch making discoveries in the history of mankind such as electromagnetic induction, magneto-optical effect, etc.

Fig. 4.50. Photograph of Faraday in 1860s.

Fig. 4.51. Photograph of Faraday later.

Faraday was given honor of being called "Prince of Science." He was innate kind and admired. The respect of people to him is larger than the honors received by him. He who believed that human could have a noble mind by researching science, contributed to development of mankind by accomplishing great works. It is not too much to say that his previous teacher Davy's most splendid discovery was finding Faraday.

6 years after he passed away, 50,000 miles of cable for telegraph was built for the first purpose. The telegraph-cable-laying ship was named "the Faraday" under permission of his wife. As the unit of capacity was named "farad," it indicated that the industry of electricity at the same time paid the highest possible compliment to him as one of the 19th century's most celebrated natural philosophers.

Chapter 5

James Clerk Maxwell

James Clerk Maxwell (1831–1879).

It was Maxwell who got Newtonian mechanics affluent, and furthermore founded electromagnetic theory theoretically unifying electromagnetic phenomena experimentally discovered by Faraday. Maxwell was regarded as the 19th-century scientist having the greatest influence on 20th-century physics.

When Einstein visited the University of Cambridge, he was told by his host that he had done great things because he stood on Newton's shoulders; Einstein replied: "No I don't. I stand on the shoulder of Maxwell."

5.1 Upbringing

On 13th June 1831, James Clerk Maxwell was born at No. 14 India Street, Edinburgh. His father John Clerk Maxwell was a laird possessing the estate of Middlebie. His mother Frances was a faithful daughter of R. H. Cay. Since James' sister Elizabeth passed away in infancy, he was only their child. When his mother passed away, leaving James motherless in his ninth year on 6th December 1839, his father Mr. Maxwell was aged fifty two. He did not marry again.

5.1.1 *Childhood at Glenlair (~10 years old)*

The part of the old estate of Middlebie inherited by John Clerk Maxwell was situated on the right or westward bank of the Water of Urr, Kirkcudbrightshire, about seven miles from Castle-Douglas, the market town, ten from Dalbeattie, with its granite quarries. It consisted chiefly of the farm of Nether Corsock, and the moor land of Little Mochrum. Before building his house, Mr. Clerk Maxwell however had added other lands to these by exchange and purchases, including the farm of Upper Glenlair.

5.1.2 *Mansion-house*

The site chosen for the house was near to the march of the original estate, where a little moor-burn from the westward fell into the Urr. The two streams of the Urr contained an angle pointing south-east. On a rising ground above the last descent towards the river, a mansion-house of solid masonry had been erected, with a pavement before the front door. On the

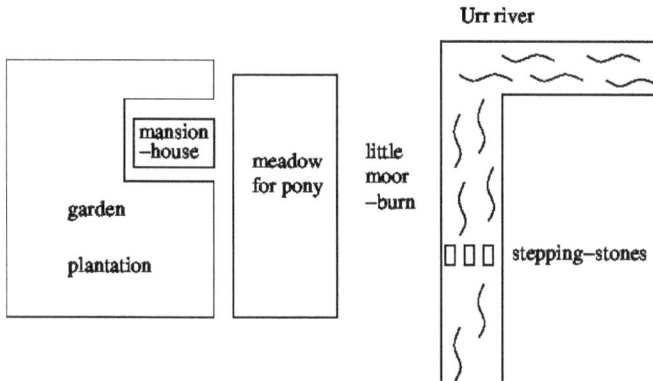

Fig. 5.1. House at Glenlair.

southwards slope, a spacious garden-ground and a plantation was beyond it, coming round to westward of the home and garden, where it ended in a shrubbery, by which the house was approached from the north. On the eastward slope, towards the Urr, there was a large undivided meadow for the ponies. To the northward, there was a yard with a duck-pond, and some humble offices or farm buildings.

At the foot of the meadow there was a ford with stepping-stones, where the bridge was afterwards to be built. At the foot of the garden a place was hollowed out for bathing. The rocky banks of the Urr, higher up, were fringed with wood, and on the upland, on either-side the moor, there were clumps of plantation, giving cover to the laird's pheasants, and breaking the force of the winds coming down from the hill of Mochrum. Glenlair was the name ultimately appropriated to the "great house" of Nether Corsock (Campbell, 1882).

5.1.3 *A child full of curiosity*

Mrs. Blackburn who was a cousin of James, told that through his infant, James always said to her, "What is it, and how does it move." Once Ms Murdoch gave a tin plate as a plaything to two years old James. On a day with a fair sky, when James set the tin plate in direction to the Sun, the light of the Sun reflected by the plate lighted up the interior of his room.

Excited James said to his nurse Maggy, "Please call Papa and Mama here." When his parents came to their son's room, their faces were lighten up. His father said, "My son, what are you doing." James replied "It is the Sun, Papa." Then, his father said, "Someday, let you view the Moon and Stars in sky." At the day with moon light later, they viewed the sky with stars at the pavement in front of the door. The infant astronomer James looked at the sky with enthusiasm.

In a woodcut drawn from the sketch of the barn ball held at Harvest-home in 1837, 6-year-old James looking at a bow of violin was expressed. Spirit similar to an acoustic discovery by Helmholtz might work on a child James. When James and his nurse returned home after walking in forest, she should keep all the collections such as a twig, a small stone and glass on a kitchen dresser until his parent would return home, in order to explain each of the collection to him. James especially was interested in a colour of stone. He also was interested in looking at the movement of insect, but he did not wounded insect. His maternal aunt Miss Cay who was a sister of his mother, was shamed that she could not answer the question of infant James.

His mother carried out the responsibility concerning education during his childhood until 1839 when she passed away. She was proud of James' good memory, because he repeated the whole of the 119th Psalm in 8 years old, and his knowledge about the Bible was extraordinary extensive and minute. He could give the chapter and verse for almost any quotation from the Psalms. His knowledge about Milton (John Milton (1608–1674: English poet)) dated from very early times. He did not memorized his knowledge by mechanically learning by heart and his knowledge occupied his imagination and sank deeper than anybody knew.

It was in Nature out of door what he was most interested in. Living thing was thing accompanying him alone, whose typical representatives were fog and tadpole. Later, he heard the discovery of Galvani concerning fog. His diversion on rainy day was to read books and sketch.

After his mother passed away, a private teacher carried out the education of James until 1841. Since James' aunt Miss Cay was aware of the relation between teacher and James being never smooth, she recommended his father that James would enroll in a school in Edinburgh (Campbell, 1882).

5.2 The Edinburgh Academy (10–13 Years Old)

On 18th November 1841, James left Glenlair for Edinburgh where there was the house of Mrs. Wedderburn (his father's sister Isabella, at the time the widow) at No. 31 Heriot Row. When he arrived at No. 31, it was snowing. A faithful servant Lizzy Mackeand also arrived together. This house would be James' living home during 8–9 years in future. Sometimes Miss Cay stayed there. His father could not leave Glenlair throughout the year, and hence he lived in two places during a year, that is, he lived at the house of his sister in Edinburgh in almost winter, living at Glenlair at the rest time.

The Edinburgh Academy was founded in 1824, being prestigious in Edinburgh. James was enrolled in the second grade. He was called by nickname "Dafty." It seemed that the boy coming from Glenlair was seen as a rustic by classmates in Edinburgh. "Daft" implied "fool." The home-room teacher A. N. Carmichael was an author of "The Edinburgh Academy Greek," being a good teacher strict with his pupils, but he could not carry out an ideal education because the junior class included 60 pupils, being too many. Therefore, James was not satisfied for the class.

At the time, James' base of living was "Old 31" house. He was given a room, where he made homework, reading books and writing poets. His cousin Thelma was learning woodcut, and James was permitted to bring out its tools. His father John if staying in Edingurgh walked with James especially on half-holiday. His father seemed an elder brother rather than his father.

5.2.1 *Field trip with his father*

The most consequential moment of Maxwell's early education occurred not in a stuffy classroom or at the mercy of a harsh disciplinarian but on a field trip with his father. In this rare outing, John took Maxwell to see a public presentation carried out by a Scottish scientist by the name of Robert Davidson. At the seminal event on February 12 in 1842, Maxwell observed Davodson's display on the subject of magnetic force and electric propulsion. This exhibit would light the spark of Maxwell's lifelong interest in electromagnetism (History, 2019).

When his father was alone in Glenlair, James sent a letter to heal father feeling lonely. The letter was written with cyphers invented by James, which were something possible to be understood by reading his diary, notebook and sketches at school life.

At the school, James was gradually finding and working his own way. He was aware of Latin being valuable for learning. He was interested in the Greek Delectus, and in biographies in the Bible. His classmate James Muirhead who would become the professor, said remembering James "Although he was a friendly boy, he never was absorbed in people surrounding him." The other classmate who would become the Rev. W. Macfarlane of Lenzie (Rev. is the title of honor for clergyman), said impression "When James enrolled in the Academy, he seemed to be a rustic and peculiar. Classmates called him with nickname "Dafty." Later James became alone."

On Sunday, he generally went to St. Andrew's Church in the morning, in the afternoon went to St. John's Episcopal Chapel by the request of Miss Cay, and became a member of Dean Ramssay's catechetical class once. Thus, James became to know both catechisms of Scotch and English Church.

5.2.2 *Polyhedrons*

It was the work carried out by James which was written in a letter for his father on his 13th birthday, what delighted him rather than bathing and writing poems. In the letter, he said "I have made a regular tetrahedron and regular dodecahedron, and two other polyhedrons, whose names I don't know." In his class, there was not a leture about Geometry not yet. He however had certainly learned the definitions in Euclid. He certainly understood the nature of the five regular polyhedrons, and constructed the five polyhedrons with accuracy.

When James returned to Glenlair after the first long stay in Edinburgh, he enjoyed the navigation at the glen of Urr in a boat as in a washtub in duck-pond at his infancy, and the shadow of his heart was softened.

5.3 Adolescence (13–16 Years Old)

The commencement of the fifth year in Academy was the time cheerful and hopeful for many pupils, because the long period of mere drill and task-work would be supposed to be over. It was the important change for Maxwell that the serious study of geometry began.

Miss Cay a sister of James' mother kept the most familiar relation among relatives with Maxwell's family, and was an aunt who was always concerned about James' future. She was about to supervise James as until then. She made an effort for him to join her companions, and to soften the peculiarity of James, letting him become a ordinary youth as the other youth. He frequently stayed in his aunt's house, where he healed her by designs and combined colours made by him. It was the first time when he applied geometry to depicting that he used accurately the perspective of a view when depicting the inner side of Roslin Chapel where his aunt was employed.

5.3.1 *To depict an oval curve*

Mr. D. Hay was a prominent decorative painter, member of the Society of Arts at the time. He intended to disregard a beauty in form and regard a mathematical principle, attracting attention among many scientists. Maxwell also was interested in his intent and discoursed on "egg-and-dart," "Greek pattern," and on the forms of Etruscan urns. It was a problem of the field of applied science how a oval curve was completely depicted. Maxwell who had learned the cross section of cone, was about to find a true practical solution of this problem.

The result researched with enthusiasm by Maxwell was published as a communication by Professor Forbes in "Proceedings of the Edinburgh Royal Society, vol. ii, pp. 89–93" (Explanation 5.1).

In the winter of 1846–47, James devoted himself to research two sciences such as magnetism and polarization of light. He also interested in Newton's rings and soap bubbles' colour change. On 25th May, he went to cutlery shop with his father in order to buy a cutter for making appropriate size of magnets.

Explanation 5.1 Oval Curves

Definition 1. If a point moves in a manner where the sum of m times its distance from one point and n times its distance from another point is equal to a constant quantity, the point will depicts an oval curve.

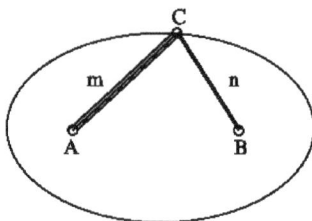

Fig. 5.2. Oval curve.

Definition 2. Two points are called foci, *m* and *n* are called powers.

Definition 3. A line combining two foci is called axis.

Problem: To depict an oval curve when foci, their powers and a constant quantity are given.

Let *A* and *B* be foci with powers *m* and *n*, and *EF* be a constant quantity. It is required to depict an oval curve. Let *C* be any point on the curve, *AC* be a distance between *A* and *C*, and *BC* be a distance between *B* and *C*. The point *C* depicts an oval curve when moving the point *C* in a manner where the relation

$$mAC + nBC = EF$$

is kept. If *m* > *n*, the focus *A* is called the greater focus and the focus *B* the less focus.

Theorem.
1. The greater focus is within an oval curve.

Proof.
If both A and B are out of an oval curve, from Fig. 5.3(a),
$EF = mAD + nBD = mAD + n(DC + BC)$ on the other hand,
$EF = mAC + nBC = m(AD + DC) + nBC$ therefore
$nDC = mDC$, $m = n$. This contradicts with $m > n$. Hence both *A* and *B* could not be out of an oval curve. □

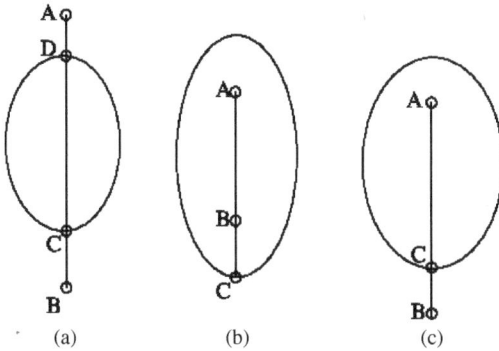

Fig. 5.3. Oval curve and position of foci.

2. If $mAB < EF$, the less focus B within an oval curve.

Proof.
It is evident that $BC = (EF - mAB)/(m + n)$. Since $mAB < EF$, $BC > 0$.
From Fig. 5.3(b), $AC = AB + BC > AB$. Hence B is in an oval curve.   □

3. If $mAB > EF$, B is without an oval curve.

Proof.
$mAB > EF = mAC + nBC > mAC$, therefore $AB > AC$.
From Fig. 5.3(c), B is without an oval curve.   □

4. If $m = n$, a curve is an ellipse.

Proof.
$EF = mAB + nBC = m(AB + BC)$. Hence $AC + BC = $ constant.
This expresses an ellipse.   □

5.4 The University of Edinburgh (16–19 Years Old)

After enrolling in the University of Edinburgh (founded in 1583), Maxwell associated with friends accompanying until that time in his original and innocent manner. His usual response was an indirect and never understandable and frequently spoken with shyness in a simple and husky tone. In his personality, he had a deep-rooted objection to the vanities and formality. He had pious fear of destroying anything — even scrapping writing paper. While in railway journeys, he traveled by the third class train with a hard seat. When at the table, he was carried away by something occurred. For example, he was fascinated by a effect of refraction of light by a finger-glass. Such a time, Miss Cay used to say "Jamsie" (a alias to James called by her), you are in a "prop" (abbreviation of mathematical proposition), and attracted his attention. He never drank wine.

5.4.1 *His acquaintances moved to University of Cambridge*

Maxwell's teacher Professor Forbes had an opinion that Maxwell had intellectual originality and enough ability. People who were aware of Maxwell's devoted affection to his father, had begun to have suggestion of his heroic devotion. His chief acquaintances were Lewis Campbell, Robert Campbell (Lewis' brother), P. G. Tait and Allan Stewart. Tait

Fig. 5.4. University of Edinburgh.

moved to Peterhouse College in University of Cambridge after one semester in 1848. Stewart moved to the same College in 1849. Maxwell however never moved until 1850.

During the three years — November 1847 to 1850, he spent impartial time in between Edinburgh and Glenlair. He studied in multifarious fields such as (1). Polarization of light (2). Galvanism (3). Rolling curves, (4). Compression of solid.

It may be guessed but cannot be known clearly why Maxwell's entrance at Cambridge had been delayed. Maxwell and his father could not part from each other because of their affectionate relation. James' delicate health would count heavily amongst the reasons, and certain floating prejudices about the "dangers of the English universities," Puseyism, infidelity, etc., had then considerable hold, especially on the Presbyterian mind (Campbell, 1882).

In April 1849, the University of Edinburgh held a meeting where Maxwell's movement to Cambridge was examined. On 1st April 1850, Maxwell got a recommendation of Professor Forbes that the University would have Maxwell move to Trinity College in Cambridge. In fact, Maxwell enrolled in Peterhouse College in Cambridge.

Fig. 5.5. Young Maxwell.

5.5 Undergraduate in the University of Cambridge (19–22 Years Old)

What Maxwell first felt in Cambridge was that he was moved to a curious city for him alone in rural area. His spirit became certainly happy even in his first term by rooted presentiment of possibility of Cambridge, together with the novelty of the scene.

But, he was anxious about his future. And this made him lend a ear to the advice for him by various quarters, that he should migrate to Trinity. The ground of the advice was that at Peterhouse the foundation was small and the chances of a fellowship there for mathematical men were less than at Trinity College. Maxwell's private tutor who was a member of Peterhouse, recommended to Maxwell's father that Maxwell should migrate to Trinity. Maxwell's own prime motive for migrating to Trinity was the hope that the larger College could afford him more abundant opportunity for self-improvement. He was not disappointed in the hope.

5.5.1 *Enrolled in Trinity College*

When Maxwell enrolled in the Trinity College, at once he made a troop of friends in multifarious fields. The experiment of Foucault's pendulum as the scientifically important event was introduced. Maxwell saw at Trinity College in April or May 1851 the experiment which proved rotation of the Earth.

When the neighborhood of Upper Corsock was lent to a hunting comrade, a member of the comrade said "It is a pity that Maxwell is not used to hunting." It was never his defect of activity but affection for animals why he was not interested in field sports. The meaning of "Wordsworth's Hart-Leap Well (William Wordsworth criticized to hunt animals for pleasure by his poem)" was Maxwell's own instinct rather than morality.

In the spring of 1852, he could have a room in the College. In April of the same year, he gained a scholarship. Hence he could concentrate attention to learning. He however impartially studied in multifarious fields. He submitted various papers to Cambridge and Dublin Mathematical Journal.

In the College life, he made himself familiar with the other scholars at the scholar's table. Especially he was about to be acquainted with scholars who were well acquainted with the Greek Roman classics.

5.5.2 *His dignity of Nazarite*

At the time his countenance, as compared with the Edinburgh days, gained in manliness, gravity and massiveness. His hair and beard were raven black, with a crisp strength in each hair, maintaining his dignity of Nazarite (though Jesus Christ was born at Bethlehem, St. Mary was in Nazareth when she heard the Annunciation to the Blessed Virgin Mary and Jesus was raised in Nazareth.) rather than 19th century youth. His plain and neat dress and an aesthetic taste gave a sublime impression together with the effect of marvelous eye.

For a short vacation term in summer of 1853, Maxwell spent at the house of Rev. C. B. Tayler, the Rector of Otley in Suffolk at east side of England. Tayler heard that Maxwell was always kind to his nephew. Maxwell for the first time felt the atmosphere of a large family which was a different from his experience raised up as one-children. When staying there, he suddenly caught a disease which was thought to be a brain fever by Tayler. Taylers tenderly nursed Maxwell as they would have nursed a son of their own. Their kindness awakened his lasting gratitude, and he referred to it long afterwards as having given him a new perception of the Love of God. One of his strong convictions thenceforward was that "Love abide, though knowledge vanish away" (Campbell, 1882).

The mathematical Contest was held in Cambridge with the result that Maxwell was the Second Wrangler, Routh of Peterhouse being the Senior and that Routh and Maxwell were declared equal as Smith's Prizemen.

5.6 Fellow of Trinity (22–24 Years Old)

Maxwell's friends in Edinburgh were satisfied with the situation that Maxwell was the Smith's Prizeman. Though Maxwell never was not interested in the success, it was his chief interest that he became free to do his life work. He was free to prepare new instruments for researching his work. He could teach some pupils by his own decision. Maxwell with his heart fully set on physical inquiries, engaged of his own accord in teaching, and undertook the task of examining Cheltenham College, and submitted to the routine which belonged to his position at Cambridge.

When actually emancipated from various works he seemed to have reverted principally at first to his Optics. He inquired into colourblind persons on all sides, and invented an instrument for inspecting the living retina, especially of dogs.

5.6.1 *Electricity*

In 1855, he read Electricity and Fluid Motion. He got on with his electrical calculations then, and worked out anything that seemed to help the understanding thereof. The research was put into shape as a paper "On Faraday's Lines of Force" in winter of 1855-6. In October 1855, he got fellowship. His name was expressed as a student amongst 3 mathematicians selected from the second year bachelors. He was appointed to lecture Hydrostatics and Optics for the three year students. Then he stopped to teach private pupils in order to keep his own research time.

For Electricity and Magnetism he took out Poisson, and began to develop more systematically his own ideas on Faraday's Lines of Force.

5.7 Professor in Aberdeen (24–25 Years Old)

Immediately after his return to Cambridge from Edinburgh where he had visited his sickly father and had confirmed for his father to be calm, Maxwell received information from Professor Forbes that the post of the Professor of Natural Philosophy became vacant in Marischal College at

Fig. 5.6. Aberdeen.

the Scottish third city Aberdeen, and that he was a candidate for the successor to the post.

He foresaw that the appointment in Scotland would satisfy his father, and that he could arrange session and vacation time to spend the whole summer uninterruptedly at Glenlair, because the distance of Glenlair from Aberdeen was nearer than the distance from Cambridge.

5.7.1 *Sickly father*

When he returned to Edinburgh about the middle of March 1856 everything was well in train. He was satisfied to know that his father also interested in the appointment of his son, and the old man regained his former vigour by it. After a few days spent in Edinburgh, the father and son went home to Glenlair, as they had planned. When, on the 2nd of April 1856, just before his son was to have returned to Cambridge, Mr. John Clerk Maxwell suddenly passed away. His candidate for Aberdeen continued. The personal loss to him was incalculable and irreparable.

After his father's death, his first duty was to apply himself to the management of his estate. He stayed at Glenlair during most of the summer, when he made a short excursion to Belfast together with his cousin, William Cay, who according to his advice was about to study Engineering under James Thomson, the brother of the Glasgow Professor William Thomson.

In November Maxwell began his work of Professor in Aberdeen. A Scotch Professor had the advantage over a College lecturer at Cambridge. If his students were less advanced, he completely directed them in his own department. Apart from any prescribed system, he could determine the order in which the parts of his subject would be developed.

As it was, his lectureship at Trinity lasted only for a year, and in the Scotch University he had only taught for three short sessions when Marischal College was on the point of being suppressed.

5.8 Marriage in Aberdeen

In the interval between his first and second sessions of 1857 at Aberdeen, he had not yet recovered from the loss of the preceding year.

In September of this year another loss renewed the feeling of desolation haunting him since his father's death. His friend Pomeroy, whom he had nursed in illness, and of whose career in India he had augured so highly, was carried off by a second attack of fever, caused by a hurried journey during the first outbreak of the Mutiny (Campbell, 1882).

His original research on electricity was now for a while interrupted by another task. In honor of the discovery of Neptune, a problem "The structure of Saturn's Ring" was set as a subject for the Adam's Prize which was given by St. John's College. The problem was extraordinary complex and could be solved by using the speculative imagination and mathematical ingenuity of prominent mathematician as Maxwell, who was completely fascinated with the problem for the time.

Amongst his new acquaintances at Aberdeen he had become most intimate with the family of Principal Dewar of Marischal College, and he frequently visited at their house. Maxwell's deep and abundant knowledge of history, literature, and theology, and the religious earnestness were there appreciated and admired.

He had been asked to join them in their annual visit to Ardhallow, the home of the Principal's son-in-law, Mr. M'Cunn, and had accepted the invitation (Campbell, 1882), spending the time with his brightest mood.

In June 1858, Maxwell got married to Katherine Mary Dewar. In May he travelled to visit his friend Campbell at Hampshire in the southern England, requesting him to attend his wedding ceremony.

He remained for two more sessions at Aberdeen. However in 1860 the College was fused, and the Professorship of Natural Philosophy at Marischal College was one of those suppressed. In the summer of 1860 the ex-professor of Aberdeen Maxwell was appointed to the vacant Professorship of Natural Philosophy in King's College, London.

5.9 King's College (29–34 Years Old)

From the time of King's College onward, Maxwell concentrated his attention on scientific research. His works in King's College were more exacting than those at Aberdeen. Nine months of lecturing in the year and evening lectures to artisans, etc., were recognized as a part of the

Fig. 5.7. King's college.

Professor's regular duties. Maxwell retained the post until the spring of 1865.

5.9.1 *Theory of gases*

In June 1860, he attended the British Association's meeting at Oxford, where he exhibited his box for mixing the colours of the spectrum. He also presented to Section A an important paper on Bernoulli's Theory of Gases; a theory which supposes that a gas consists of a number of independent particles moving about among one another without mutual interference, except when they come into collision. Maxwell showed that this theory could satisfactorily explain the viscosity of gases, their low conductivity for heat, and Graham's laws of diffusion.

On the 17th May 1861, he lectured "On the Theory of the Three Primary Colours" for the first time before the Royal Institution.

5.9.2 *The research on electricity and magnetism*

All this while at King's College, Maxwell was quietly and securely laying the foundations of his great research on Electricity and Magnetism, for bringing it to completion.

In 1862–63, the experimental measurements by which the present standard of electrical resistance (the Ohm) was first determined, were made at King's College by a sub-committee of the B. A. (British Association), consisting of Maxwell, Balfour Stewart, and Fleeming Jenkin, in accordance with a method proposed by Sir William Thomson. In 1881, an International Commission made a determination of the standard of resistance first measured by the B. A. Committee.

Another important experimental investigation conducted by Maxwell about this period was the determination of the ratio of the electromagnetic and electrostatic units of electricity, for the purpose of comparing this quantity with the velocity of light. The experiment amounted to a comparison between the attractions of two electric currents flowing in coils of wire, and the attraction or repulsion between two metal plates which have each received a charge of electricity. Maxwell had pointed out that, in accordance with his theory, the ratio of the units should be equal to the velocity of light (Campbell, 1882).

5.9.3 *Excellent sick-nurse*

During his residence in London his brother-in-law, the Rev. Donald Dewar, came and stayed in his house in order to undergo a painful operation. Maxwell offered the ground floor of his house to Mr. Dewar and his nurse. He himself, meanwhile, used to take his meals in a very small back room, where frequently he breakfasted on his knees. Maxwell seemed to act frequently in the capacity of nurse to Mr. Dewar, who would always look out anxiously for his return from College, and whose face would light up with a smile of pleasure and relief when he saw him coming (Campbell, 1882).

It was the pleasant incident during his staying in London that he could improve his acquaintance with Faraday, with whom he seemed to have dined on the occasion of his lecture before the Royal Institution in 1861.

5.9.4 *Lecture at royal institution*

On one occasion he was wedged in a crowd attempting to escape from the lecture theater of the Royal Institution, when he was perceived by Faraday, who alluding to Maxwell's theory of Gases consisting of a number of molecules, accosted him in this wise- "Ho, Maxwell, cannot you get out? If any man can find his way through a crowd it should be you" (Campbell, 1882).

At the beginning and at the close of the King's College period Maxwell suffered from two severe illnesses, which were both dangerously infectious. In both of them, he was nursed by Mrs. Maxwell. When he went to the fair at Glenlair in order to buy a pony "Charlie" for his wife, he was supposed to have caught smallpox. During this illness, his wife was left quite alone with him — the servants only coming to the door of the sick-room, because of infectious nature of the disease. The servants heard him say that by her assiduous nursing on this occasion she saved his life.

The second illness occurred in September 1865, also at Glenlair, where Maxwell had been riding a horse, and got a scratch on the head from a bough of a tree, followed by an attack of erysipelas. Mrs. Maxwell was again his nurse.

5.9.5 *At Glenlair*

Afterwards he resigned his post at King's College, and spent the most part of years followed the resignation, at Glenlair. He embodied some of the results of his investigations in substantive books, taking advantage of this retirement. Although not published till 1873, the great work on Electricity and Magnetism was now taking definite shape, and the treatise on Heat, which appeared in 1870, had been undertaken as a by-work during the same period.

At Glenlair, Maxwell had an affectionate relationship with his neighbors and with their children. He used to occasionally to visit any sick person in the village, and read and prayed with them in cases where such ministrations were welcomed.

Maxwell's retirement was not by any means unbroken. In the years 1866, 1867, 1869 and 1878, he was either Moderator or Examiner in the Mathematical Tripos at Cambridge, where his influence was more and more felt. His work on these occasions was a principal factor in the movement, which led ultimately to important changes in the Examinations system; to the creation of the Cavendish Laboratory, and to the foundation of the Chair of Experimental Physics (Campbell, 1882).

His paper on the Viscosity of Gases, printed in the Philosophical Transaction for 1866, had been delivered by him as the Bakerian Lecture for that year.

5.10 Cambridge (1871–1879)

On the 9th February 1871, the Senate of the University of Cambridge founded the Chair of Experimental Physics, the new professorship. The Duke of Devonshire, who was Chancellor of the University, had signified his desire to build and furnish a Physical Laboratory for Cambridge in October 1870, perceiving how useful such an institution might be made, as a member of the Royal Commission on Scientific Education.

It was the question with anxiety for some time who should be the first professor. It was understood that Sir William Thomson had declined to stand, and it was thought uncertain whether Maxwell could be persuaded to leave the retirement of his country-seat. At first Maxwell hesitated to become a candidate, from genuine diffidence, and then he was induced to do so, on the understanding that he might retire at the end of a year, if wishing. On the 24th February, his candidacy was announced. And he was appointed on the 8th March with no opposition.

In the Senate of 9th February, it had been enacted that "the principal duty of the professor of Experimental Physics" should be to teach and illustrate the laws of Heat, Electricity and Magnetism; to apply himself to the advancement of the knowledge of such subjects; and to promote their study in the University.

5.10.1 *The Chair of Cavendish Laboratory*

For some time after his appointment, Maxwell's principal work was to design and to superintend the erection of the Cavendish Laboratory.

He inspected the Physical Laboratories of Sir William Thomson at Glasgow and of Professor Clifton at Oxford, in order to embody the best features of both of these institutions in the new Laboratory. However many of the most important arrangements were of his own invention. An account of the Laboratory itself will be found in *Nature* (vol. x. p. 139). The architect Mr. W. M. Fawcett of Cambridge appeared to have fully appreciated and thoroughly carried out all Professor Maxwell's suggestions.

Even in 1874, however manifold apparatuses were still desired, and the Duke expressed his wish to furnish the Laboratory completely with the necessary apparatuses. In the annual report to the University in 1877 Professor Maxwell announced that the Chancellor had now completed his gift to the University, by furnishing the Cavendish Laboratory with apparatus suited to the present state of science; but at the same time he wrote to the Vice-Chancellor stating that he should reserve to himself the privilege of presenting to the Laboratory such apparatus as the advancement of science might render it desirable for the University to possess. And during the short remainder of his tenure of the professorship he expended many hundreds of pounds in this manner. The Cavendish Laboratory opened in 1874 (in 1974 it moved to the present place in Westside of Cambridge).

5.10.2 *Verifying Ohm's law*

Mr. Chrystal was undertook a series of experiments for verifying Ohm's Law respecting the relation between the current and the electro-motive force in a wire, on which some doubt had been thrown by Weber's theories, and in an opposite direction, by a series of experiments reported to the British Association by Dr. Schuster in 1874.

In consequence of these doubts a committee was appointed by the British Association consisting of Professor Maxwell, Professor Everitt, and Dr. Schuster, and the report of this committee was presented to the Association at their annual meeting in Glasgow in 1876. According to the report, the investigation proved that when a unit current passes through a conductor of a square centimetre section, its resistance does not differ from its value for indefinitely small currents by 0.000,000,001 percent.

Maxwell still found occasional pleasure in riding at Cambridge as well as more frequently at Glenlair, where he resided as much as his

Fig. 5.8.   Cavendish laboratory (at the opening).

Fig. 5.9.   James Clerk Maxwell (in 1874).

behaviour could be consistent with his professional duties. He always arranged to leave Cambridge at the end of the Easter term in time to offici- ate at the midsummer communion in the Kirk at Parton, where he was an elder.

His liberality in his own neibourhood was very great, because besides the endowment of the church and building of the manse at Corsock, he had planned a large contribution to cause of primary education.

During the last few years of Maxwell's life, Mrs. Maxwell suffered from the serious and protracted illness. He was an excellent sick-nurse, as we have already seen how he devoted himself to his father during his ill- ness, and how he cured for his brother-in-law when in London. On one occasion during Mrs. Maxwell's illness he did not sleep in a bed for three weeks, but conducted his lectures and other work at the Laboratory as usual.

In November 1876, he was elected a member of the Council of the Senate of the University. He was also a member of the Mathematical Studies and Examinations Syndicate, which was appointed on the 17th May 1877, and which sat every week during term for a whole year for the purpose of reorganizing the Mathematics Tripos. He was the president of the Cambridge Philosophical Society during the session 1876-7.

Although the publication of the Treatise of Heat and of the Electricity and Magnetism falls within this period, they were mainly written during the time of his retirement at Glenlair.

5.11 Illness and the Last (1879–48 Years Old)

After his recovery from the attack of erysipelas at Glenlair in 1865, Maxwell had been fairly good until the spring of 1877. He then began to be troubled with dyspeptic symptoms. One day in 1877, on coming into the Laboratory after his luncheon, he dissolved a crystal of carbonate of soda in a small beaker of water, and drank it off. A little while after this he felt a painful chocking sensation. In June 1877 when he returned, as usual, to Glenlair, he said that he felt like a child, as for some time he had been allowed no food but milk. The definitive prognosis would not come until a certain Professor Sanders paid Maxwell a visit on 2nd October

1879. Upon his examination, Sanders quickly determined that Maxwell was indeed suffering from abdominal cancer and, not attempting to sugar-coat the facts, bluntly informed him that he most likely only had a few weeks left of life. Maxwell's care at this point turned from seeking a remedy to seeking mere mitigation of discomfort (History, 2019).

On the Saturday preceding his death he received the Sacrament of the Lord's Supper from Dr. Guillemard, and it was while Dr. G. was putting on his surplice that Maxwell repeated to him George Herbert's lines on the priest's vestments, entitled Aaron. Maxwell's mind and memory remained perfectly clear to the very last (Campbell, 1882). On the 5th November 1879 he passed away in Cambridge, 48 years old.

There was a preliminary funeral ceremony in Trinity College Chapel, where the first part of the Burial Service was read, in the presence of all the leading members of the University. The body was then taken home to Glenlair, and buried in Parton Churchyard, the funeral being attended by number of his countrymen from far and near.

The leading note of Maxwell's character is as follows. His tenderness for all living things was deep and instinctive; from earliest childhood he

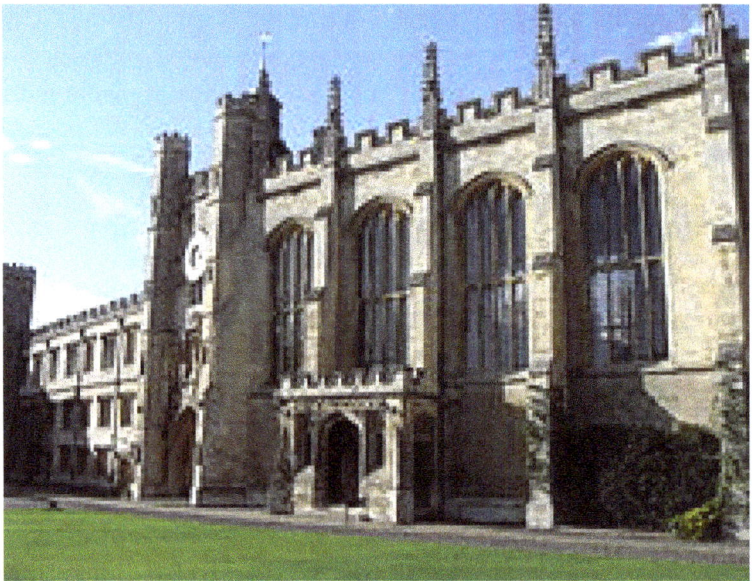

Fig. 5.10.   Trinity College Chapel.

could not hurt a fly. Not less instinctive was the sense of equality amongst all human beings.

His aunt, Mrs. Wedderburn, who had had the care of him during so much of his early life, said on the occasion of his marriage, "James has lived hitherto at the gate of heaven."

Maxwell inherited Newton, and developed "Kinetic gas theory," yielding affluent Newtonian mechanics. Furthermore, he inherited Faraday, and theoretically unified electromagnetic phenomena experimentally discovered by Faraday, having laid the foundation of Electromagnetism. His success had contributed together with Newtonian Mechanics to development of modern physics (quantum mechanics and relativistic theory) in the beginning of 20th century.

Appendix 5.1 Maxwell's Concept on Electricity and Magnetism (Yukawa & Tamura, 1955–62:II)

Line of force

After taking his Bachelor degree in 1854, Maxwell read through Faraday's Experimental researches. In Faraday he found a mind essentially of his own type. Thoroughly conversant himself with the theory in Mathematical Treatises, Maxwell saw the connection between Faraday's point of view and the method of research adopted by the Mathematicians, though Faraday had not use mathematical expression. He began to embody this understanding mathematically as paper "On Faraday's Lines of Force" (Maxwell, 1855–56).

For the purpose of getting physical idea without taking physical theory, it is necessary to be well versed in a similarity in physics. Maxwell tried to present mathematical conception necessary for researching electrical phenomena using such similarity, intending to mathematically express Faraday's point of view using similarity. At first, he began to express an idea of line of force on the basis of similarity. Maxwell considered a similarity between static-electricity and hydrostatics, where the hydrostatics was an ideal fluid. He completed similarity between electric force and velocity of fluid, similarity between potential difference in electricity and pressure difference in fluid, similarity between line of force in electricity and line of fluid, similarity between tube of force in electricity and tube of fluid and similarity between sphere with equal potential and sphere with equal pressure in fluid, etc. He explained laws of static-electricity on the basis of these similarity, furthermore, explained Ampere's law. He tried to discuss on electromagnetic induction, but could not explain it on the basis of similarity.

Similarity between static-electricity and hydrostatics

His first treatise (1855) was the paper which only translated Faraday's point of view to mathematical words on the basis of similarity between static-electricity and hydrostatics. However, in the second treatise "On Physical Lines of Force (Maxwell,1861–62)," he planned to theoretically unify all the electromagnetic phenomena on the basis of similarities furthermore developed.

For example, Faraday's one electric line of force $\mathbf{D}$ starts from unit electric charge as Fig. 5.11. Hence the quantity of electric lines of force is equal to the quantity of electric charge. This description is translated to mathematical description as follows: the quantity of electric lines of force $\mathbf{D}$, is expressed as spherical integral $\int(\mathbf{D} \cdot d\mathbf{S})$, where $d\mathbf{S}$ is the segment vector of surface $\mathbf{S}$ of sphere containing electric charge, and $(\mathbf{D} \cdot d\mathbf{S})$ is scalar product of $\mathbf{D}$ and $d\mathbf{S}$, meaning the quantity of $\mathbf{D}$ through $d\mathbf{S}$. The quantity of electric charge is expressed as volume integral $\int \rho \, dV$ where ρ is the density of electric charge and V is the volume containing the charge. Hence we have $\int(\mathbf{D} \cdot d\mathbf{S}) = \int \rho dV$. By Gauss' theorem we have

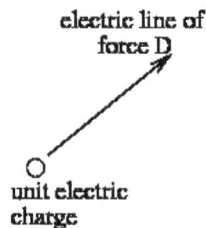

Fig. 5.11.   Faraday's electric line of force.

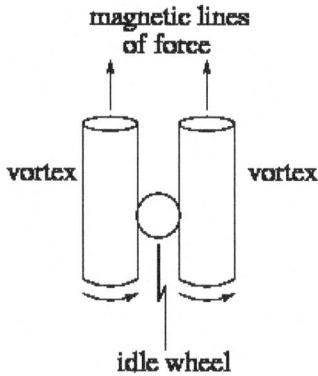

Fig. 5.12. Vortex and idle wheel.

$\int \mathrm{div}\, \mathbf{D}\, dV = \int \rho dV$. Therefore, we have the third equation among Maxwell's electromagnetic equations as mathematical translation:

$$\mathrm{div}\, \mathbf{D} = \rho \text{ (Explanation 5.2)}.$$

Examining electromagnetic phenomena from the point of view of mechanics, Maxwell had a conception that there existed magnetic influence in a medium as a tension or pressure, and two types of forces in fluid were a tension along with an axis and an equal hydrostatic pressure perpendicular to the axis. From the relation between magnetic lines of force and attractive or repulsive forces between magnetic poles, Maxwell considered that the force along with axis of magnetic line of force was tension.

Maxwell's vortex model

Maxwell set a similarity between a magnetic field and medium having vortex-group structure, where the pressure perpendicular to line of force arose from centrifugal force of vortex rotating around the axis of line of force, and a magnetic force was expressed as the rotating velocity of vortex.

In this Maxwell's model, all the vortexes in medium should rotate in the same direction with the same rotating velocity in the case of uniform magnetic field, where neighboring vortexes moving in opposite direction on the contact side. It was the problem how such vortex-group structure was constructed. The problem was solved by using an idle wheel, which was set between neighboring vortexes, and had a one-layer with particles (Fig. 5.12). Particles in a idle wheel had not the friction each other, but had the friction against vortexes. Maxwell called the particles in idle wheel "electric particles," and thought that electric current was formed by the displacement movement of the particles. The tangential pressure received by particles in wheel from vortex, corresponded to "inductive electromotive force," and the pressure between each other particle corresponded to the "voltage." He showed how laws of static current and electromagnetic induction were derived from such model.

Next the vortex model was applied to research on electro-static phenomena by using elasticity of vortex. It was necessary to understand how a material charged influenced neighboring medium, that is, to elucidate discrimination between conductor and insulator by using model. In the case of conductor, "electric particle" was free to move from molecule to other molecule — vortex is thought to be smaller than material molecule —. On the other hand, in the case of insulator, electric particle only displaces within a molecule. As long as this displacement occurs, current called "Maxwell's displacement current" occurs. This displacement occurs until the elastic reaction of vortex holds equilibrium with electric force causing the displacement moving, and when electric force removed the vortex recovered to the pre-position by elasticity. That is, conductor has a similarity to viscous fluid for idle wheel, and insulator has a similarity to elastic solid. Thus, consideration proceeded to the relation between the displacement and the force arising from the displacement, and electrostatic phenomena was developed elastically.

It was important that Maxwell's consideration with model could not only derive the conventionally known laws including electromagnetic equations, but also could predict a new phenomenon. In the medium thus modeled, it is inevitably derived that periodic vibration of displacement may occur and transmits as a wave.

Prediction of electromagnetic wave being equivalent to light

Maxwell found that the velocity of the electromagnetic wave predicted by him was equal within experimental error range to the velocity of light discovered by Fizeau, using experimental measurements of Rudolph Kohlrausch (1809–58) and Weber (1856). And he concluded that light is a wave vibrating in a direction perpendicular to the moving direction which arises from electromagnetic phenomenon, that is, he predicted that electromagnetic wave was equivalent to light.

Maxwell's ultimate thought

However Maxwell thought that vortex-group construction was only a model, and never expressed a real situation. His concept of vortex-group construction was only used for the purpose that he estimated similarity and expressed the relations between the electromagnetic phenomena and mechanics. He thought that the concept should not be expected to be more than that above mentioned. His thought had been embodied in the Treatise "A Dynamical Theory of the Electromagnetic Field" (1864) and Treatise "On Electricity and Magnetism" (1873).

Explanation 5.2 Four Electromagnetic Equations

In 1873, Maxwell derived electromagnetic equations described by four partial differential equations which theoretically unify all the electromagnetic phenomena, where the partial differential is the differentiation of a multivariable function with respect to one variable. Although in Explanation 4.7 we explained a part of electromagnetic equations, in this explanation we explain in detail those. Four electromagnetic equations are given as follows:

The first equation: electromagnetic induction;
Left-hand side = rot $\mathbf{E}$ = $(\partial E_z/\partial y - \partial E_y/\partial z, \partial E_x/\partial z - \partial E_z/\partial x, \partial E_y/\partial x - \partial E_x/\partial y)$ (Explanation 4.7). The electric field $\mathbf{E}$ is a force acting to unit electric charge; spatial gradient of voltage.

Right-hand side = $-\partial\mathbf{B}/\partial t$: the negative value of temporal change in the magnetic flux density $\mathbf{B}$. $\mathbf{B} = \mu\mathbf{H}$, where $\mathbf{H}$ = magnetic field and μ = magnetic permeability. $\mathbf{B}$ is also called magnetic induction.

The second equation: the occurrence of a magnetic field $\mathbf{H}$ due to current and a temporal change in the electrical displacement $\mathbf{D}$, where $\mathbf{D} = \varepsilon\mathbf{E}$, ε = dielectric constant.
Left-hand side = rot $\mathbf{H}$.
Right-hand side = $\mathbf{i} + \partial\mathbf{D}/\partial t$, $\mathbf{i}$: current through unit cross section, $\partial\mathbf{D}/\partial t$: temporal change in $\mathbf{D}$. The temporal change in $\mathbf{D}$ is called displacement current.

The third equation: the relation between electrical displacement $\mathbf{D}$ and charge.
Left-hand side = div $\mathbf{D} = \partial D_x/\partial x + \partial D_y/\partial y + \partial D_z/\partial z$, D_x: x-component of $\mathbf{D}$.
Right-hand side = ρ: quantity of electric charge per unit volume.

The forth equation: the equation of continuity of magnetic flux density $\mathbf{B}$.
Left-hand side = div $\mathbf{B}$.
Right-hand side = 0
    The curves expressing directions of electrical displacement $\mathbf{D}$ and magnetic flux density $\mathbf{B}$ are called "electric line of force" and "magnetic line of force," respectively.

Meaning of equations
The meaning of each equation is explained in the following.

The meaning of the first equation:
At first, we integrate both sides of the first equation over area within a coil in electric circuit. Then, the left hand side becomes a integration of rot $\mathbf{E}$ over area within a circle of coil, and becomes an integration of $\mathbf{E}$ along a line of coil, which expresses an electromotive force (voltage), because $\mathbf{E}$ being the spatial gradient of voltage, by Stokes' theorem meaning that an integration of rot $\mathbf{E}$ over area becomes an integration along a line of boundary of the area. On the other hand, right hand side becomes a temporal change of total quantity of $\mathbf{B}$ (that is total number of magnetic lines of force). Hence the first equation means that the temporal change of $\mathbf{B}$ is equal to inductive electric motive force, that is, electromagnetic induction

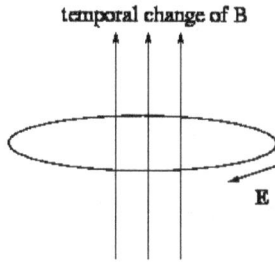

Fig. 5.13. Electromagnetic induction.

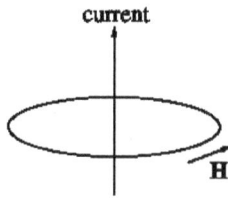

Fig. 5.14. Magnetic field due to current.

(Fig. 5.13). Induced current yields reverse magnetism to temporal change of **B**.

The meaning of the second equation:
When we integrate both sides of the second equation within a circle with radius r, left-hand side becomes an integral of magnetic field **H** along circle line. Right-hand side becomes current through the circle with radius r. Hence the second equation means the occurrence of magnetic field due to current, that was discovered by Oersted (Fig. 5.14).

The meaning of the third equation:
When we integrate both sides of the third equation within a volume, left-hand side becomes the sum of number of electric lines of force coming out from the boundary surface of the volume, by Gauss' theorem meaning that the volume integral of div **D** is equal to the surface integral of **D** on the boundary surface of the volume. Right-hand side becomes the total quantity of electric charge. Hence the third equation means that the total number of electric lines of force is equal to the total quantity of electric charge (Fig. 5.15).

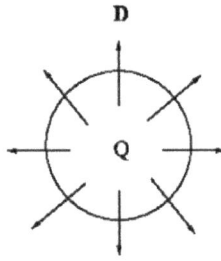

Fig. 5.15.   D and charge Q.

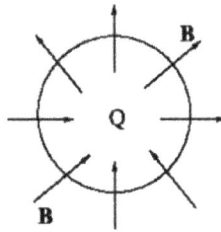

Fig. 5.16.   Continuity of **B**.

The meaning of the forth equation:

When we integrate both sides of the forth equation within a volume, left hand side becomes the total number of magnetic lines of force coming out from the boundary surface of the volume. Since right hand side is equal to zero, it means that the number of magnetic lines of force coming out from the surface is equal to the number of magnetic lines of force coming into the surface. That is, all the magnetic lines of force coming into the volume should come out from the volume, meaning the continuity of magnetic line of force (Fig. 5.16).

Explanation 5.3 Ampere's Law

When the current I flowes on the straight line of wire, the magnetic field H on the circle with radius r occurring due to I is given by the value of I divided by 2πr.

Explanation 5.4 Graham's Law

Graham experimentally discovered in 1846 that the proliferation of gas is inversely proportional to the square root of its mass, meaning that the smaller gas proliferates the more.

Explanation 5.5 Maxwell's Kinetic Gas Theory (Yukawa & Tamura, 1955–62:I)

The first and second laws of thermodynamics

All the systems of thermodynamics hold entropy S which is determined only by the state of the system. When the system is changed gradually keeping thermal equilibrium (called quasi-static changing) from starting state A to final state B, the change dS of entropy S during the process from *A* to *B* is given by the sum dQ of little heat absorbed during the process, divided by the absolute temperature T, yielding

$$dQ = T \, dS.$$

The process satisfying this equation is called reversible process. In case of irreversible process, the following relation is satisfied:

$$T \, dS > dQ.$$

Especially in adiabatic process, $dS > 0$.

Since Rudolf Clausius succeeded to formulate the second law of thermodynamics (law of increase of entropy: when the heat transfers from a body with high temperature to a body with low temperature, and also does not leave any change to nature, the process of heat transfer is irreversible process, and entropy increases.), the theory of heat could be discussed systematically as the thermodynamics, together with the first law of the thermodynamics (inside energy = heat dQ given from outside — the work dA acted to outside).

However, the development of thermodynamics relating the spatial movement of molecules to thermal phenomena on the basis of atomic structure was not yet carried out. This development was at first carried out by *D*. Bernoulli in Hydrodynamica (1738), who derived law of Boyle concerning the relation between pressure and volume, considering the elasticity of gas being due to collision of molecules against vessel.

When Joule (1848) and Kronig (1856) tried to derive the state equation of gas and specific heat based on the molecular theory of heat, they thought that any molecular movement could not be calculated because of extreme irregularity, but it could be treated by probabilistic calculation.

Probabilistic calculation by Maxwell

It was J. C. Maxwell who carried out the probabilistic calculation on the molecular theory of heat. He introduced the concept of "mean free path" which is given by the average of velocity of molecule divided by the number of collisions per unit time, and is inversely proportional to the density of gas, and set the concept of number of collisions as the basis of molecular theory.

In the case of general molecular group rather than ideal gas, Maxwell considered the distribution of velocity of molecule among the uniform system in thermal equilibrium, under the condition of isotropic property and spatial uniformity in the velocity space.

When the probabilities that three independent velocity components in the velocity space take the value (ξ, η, ζ), are given by $w(\xi)$, $w(\eta)$, $w(\zeta)$, the probability $W(v)$ that molecular velocity takes the value v whose components are ξ, η, ζ, is given by their product (Yukawa & Tamura, 1955–62:I453).

$$W(v) = w(\xi)w(\eta)w(\zeta), \tag{1}$$

Taking natural log and differentiating with respect to ξ, we have

$$(1/v)[dW(v)/dv]/W(v) = (1/\xi)[dw(\xi)/d\xi]/w(\xi) \tag{2}$$

where $v = (\xi^2 + \eta^2 + \zeta^2)^{1/2}$. (2) is constant because derivative of (2) with respect to η or ζ is 0. Let this constant be -2β, we have

$$(d/d\xi)\ln w(\xi) = -2\beta\xi$$
$$w(\xi) = a \exp(-\beta\xi^2) \tag{3}$$
$$\int w(\xi)\, d\xi = a \int \exp(-\beta\xi^2)\, d\xi = 1$$

Since $\int \exp(-\beta\xi^2)\, d\xi = (\pi/\beta)^{1/2}$, hence

$$a = (\beta/\pi)^{1/2}$$

From the law of equipartition of energy,

$$(1/2)\mu\langle\xi^2\rangle = (1/2)\mu\langle\eta^2\rangle = (1/2)\mu\langle\zeta^2\rangle = (1/2)k_B T \tag{4}$$

where $\langle\cdot\rangle$ means average, k_B the Boltzmann constant, T absolute temperature, and μ the mass of a molecule.

By probabilistic calculation of $\langle\xi^2\rangle = \int \xi^2 w(\xi)\, d\xi$ using partial integral formula, we have

$$(1/2)\mu\langle\xi^2\rangle = \mu/(4\beta) \tag{5}$$

Maxwell's velocity distribution law
From (4) and (5),

$$\beta = \mu/(2k_B T) \tag{6}$$

$$W(v) = [\mu/(2\pi k_B T)]^{3/2} \exp[-E/(k_B T)], \tag{7}$$

$$E = (\mu/2)v^2 = (\mu/2)(\xi^2 + \eta^2 + \zeta^2)$$

$W(v)$ is probability for one molecule, hence in case of considering unit spatial volume, for the density n we have the probability $f(v)$:

$$f(v) = n[\mu/(2\pi k_B T)]^{3/2} \exp[-E/(k_B T)] \tag{8}$$

In the case of considering the absolute value of v, we should consider the area of surface of sphere with radius v in velocity space, hence the probability for the absolute velocity value being v is given by

$$F(v) = 4\pi v^2 f(v) = 4\pi n[\mu/(2\pi k_B T)]^{3/2} v^2 \exp[-E/(k_B T)] \tag{9}$$

$$E = (1/2)\mu v^2$$

$F(v)$ includes the term: $v^2 \exp[-E/(k_B T)]$ which is the product of the exponential function of variable $(-v^2)$ and v^2. v^2 is the function increasing with

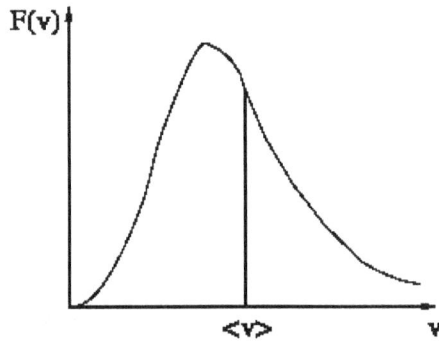

Fig. 5.17. Maxwell's velocity distribution $F(v)$, v; velocity of molecule, $\langle v \rangle$; average velocity.

v and $\exp[-E/(k_B T)]$ is the function decreasing with v. Hence $F(v)$ varies with v as Fig. 5.17.

Thus, he derived a famous Maxwell's velocity distribution law (1859), showing capability of statistical average. This distribution was a standing point of development of kinetic theory of gas.

Maxwell's theory was powerfully developed by Ludwig Boltzmann (1844–1904). Boltzmann tried to elucidate the irreversibility — proposition of entropy increase — which is characteristic to thermal phenomenon on the basis of thermal molecular theory, and expressed it as H-theorem (1871, 1872). He considered that characteristic of the irreversibility is statistical — dynamical. This means that the elemental characteristic of molecular group is not deterministic, but statistical — probabilistic. Hence kinetic theory of gas was not a simple extension of Newtonian mechanics, but meant a new expansion. Thus, it was requested to develop "statistical dynamics," which had proceeded from Boltzmann's effort.

Chapter 6

Albert Einstein

Albert Einstein (1879–1955) (sketch by author).

Albert Einstein founded the relativistic theory, deriving the new conception on the time-space and introducing revolution to physics. The relativistic theory constituted with quantum mechanics the two greatest theories of modern physics. The relativistic theory played a powerful guideline on developing cosmology. Among scientists at the beginning of the 20th century, anyone could not make more contribution to epoch-making progress in physics than Einstein.

6.1 Upbringing

6.1.1 *Birth of Einstein*

On 14th March 1879, Einstein was born at Ulm in southern Germany. His parents were Jews, but they were not enthusiastic Jews, and did not perform Jewish rites. His father Hermann (Hermann Einstein) was calm, kind and loved by all acquaintances. He liked literature and read Schiller and Heine to family. He had small electrical construction business, but was in financial difficulties. When Einstein was at the age of 1, family relocated

Fig. 6.1.   Einstein and his sister Maria (photograph in 1885).

Fig. 6.2. Ulm Cathedral (sketch by author).

to Munich. On 18th November 1881, his sister Maria (always called Maja) was born. Maja was to become her brother's most intimate soul mate (Isaacson, 2017:11).

In Munich father with uncle Yakob started enterprise for facilities of gas and tap water. In 1885, they started electro-technical factory to produce Dynamo, arc lamps, and electrical measuring equipment for municipal electric power stations and lighting systems. The fund was invested by mother's father. Einstein and his sister liked the life of house with large garden with large trees.

6.1.2 *Unusually long time before he could speak*

Einstein's head was unusually large as the baby's head. He liked being alone. He was not interested in game and toy which children liked. The child might be backward because of unusually long time before he could speak. Whenever he had something to say, he would try it out on himself, whispering it softly until it sounded good enough to pronounce aloud. His younger sister recalled "no matter how routine, he repeated to himself softly, moving his lips" (Isaacson, 2017:8). His mother worried whether he was under poor condition. But because the reason was known, it was found that there was no anxiety. The reason was that his thought was internally profound, and before uttering words, he considered thoroughly in brain. Hence he was taciturn and very quiet boy. At the age of 5, to please him in bed due to disease, his father gave a small compass. The compass impressed so strongly that he trembled. "Behind magnetic needle there should be something laying deeply hidden," he said afterward about the experience at his very early in life (Pais, 1982:37).

Competent pianist mother Pauline decided to introduce musical education to her children. At the age of 6, Einstein took violin instruction, and his sister learned piano. He disliked repetition of practice, and was not interested in violin. After he heard sonata of Mozart, music became important to him, and after coming of age, his performance of violin became skillful. Music was not only diversion but also important aid. On the contrary, it helped him to think (Isaacson, 2017:14). That is, when he faced a difficult challenge in his work, playing violin solved all his difficulties (Isaacson, 2017:14).

6.1.3 *Entering public school*

At the same age of 6, Einstein entered public school, the Volksschule. He was so superior that his mother said to grandmother "Albert is again excellent Grade." He was patient student and solved mathematics with confidence. Because the play which he liked necessitated patience and persistence, he did not play with classmate, and was gentle boy.

At the age of 7, uncle Yakob started to teach algebra him. Feeling anxiety that problem was difficult for him, Yakob gave difficult problems continuously, but Einstein always solved problems and was immersed in delight (Isaacson, 2017:17). In October 1888, He entered the Luitpold Gymnasium. He was always the most superior, and especially superior in mathematics and Latin. However, he was not satisfied with the school. Because in the school there were authoritarian teachers and servile students, and rote learning was performed.

Fig. 6.3.   Einstein at the age of 14 (photograph in 1893).

A medical student Talmud (Max Talmud) with little money came for dinner at every Thursday night. After dinner, he discussed with Einstein on science and philosophy, and gave him scientific book, and important educational influence.

In 1894, father failed in enterprise, and planned to move factory to Italy according to recommendation of Italian Garrone (Signor Garrone) comrade of enterprise. Family relocated to Milan, but he stayed in Munich to complete schooling. In 1895, because of completion of factory, family again relocated to Pavia.

Einstein staying in Munich was depressing, and missed his family. He disliked school. Without consulting parents he decided to go to Italy. Requesting doctor to prepare a medical certification of poor physical condition, he took a leave of absence from the Gymnasium, and in 1895 he relocated to Pavia. To parents who were astonished by his sudden arrival, he informed that "afterward, he self-studies and tries the entrance examination of ETH (Eidgenossische Technische Hochschule Zurich: ETH Zurich)" (Pais, 1982:40). He disliked authoritarian education in the Gymnasium, and halfway left the school. Furthermore he decided to give up his German citizenship. Italian landscape and the arts impressed profoundly him who started his new life (Pais, 1982:40). He who was quiet, changed to be lively and talkative youth.

6.2 Eidgenossische Technische Hochschule Zurich: ETH Zurich

6.2.1 *Entrance examination for ETH*

In October at the age of 16, Einstein took an entrance examination of ETH for studying electrical engineering, but he failed in the examination. As previously decided, he gave up his German citizenship. Afterward, for some years, he would not get any citizenship.

For the purpose of preparing again entrance examination of ETH, he went to the cantonal school in Aarau in German area in Swiss. He boarded at the house of Winteler, a teacher of the school. The family Winteler were good people. There was free atmosphere in the school, and he could respect teachers, and he could enjoy school life. The scars of failure in entrance examination disappeared in mind. He thought that if he would

Fig. 6.4. ETH Dome (sketch by author).

pass the entrance examination, he would go to Zurich. For 4 years there, he would study on mathematics and physics. In future, he would be a teacher of science especially theoretical field. Because he was lack ability of practical field, abstract mathematical thinking fitted his temperament (Pais, 1982:40).

In 1896, the factory possessed by his father and uncle was in financial difficulties and bankrupt. Uncle found a work in large company, but his father intended to start a new factory. Einstein warned his father not to start new enterprise, and visited uncle to request him not to support his father. He sent to his sister a letter describing "parent's often misfortune weighs heavily on my mind. I as adult am a burden of family, and it anguishes me that I cannot support parents" (Pais, 1982:41). Two years later, his father found new work in electric company. Einstein's depression ended by this.

6.2.2 *Study by himself*

On 29th October 1896, at the age of 17, Einstein passed the entrance examination for ETH. Upon satisfactory completion of the four-year curriculum, he would qualify as a Fachlehrer, a specialized teacher in mathematics and physics at a high school. He did not attend lectures, and studied by himself researches of Kirchhoff (Gustav Kirchhoff), Hertz (Heirich Rudolph Hertz), Helmholtz (Hermann Ludwig Ferdinand von Helmholtz) and electromagnetic theory of Maxwell. He considered Minkowski (Hermann Minkowski) to be superior teacher of mathematics, but he did not attend his lecture.

Without relying only lectures of university, he expanded field of academy with wide view. He read papers of Lorenz (Hendrik Antoon Lorentz) and Boltzmann (Ludwig Eduard Boltzmann). In Zurich he got acquainted with many friends, and lived pleasant student life. 4 years later, in 1900 at the age of 21, Einstein became qualified as a Fachlehrer. The maximum in the examination was 6.0, and his final grades were 5.0 for theoretical physics, 5.0 for experimental physics, 5.0 for astronomy, 5.5 for function theory and 4.5 for short paper on thermal conduction (Pais, 1982:44–45). Because he did not attend lectures, he prepared for the examination borrowing notebook from Grossmann (Marcel Grossmann). In the same year,

Planck (Max Karl Ernst Ludwig Planck) proposed the concept of "energy quantum" (Appendix 3.1).

6.3 The Patent Office in Bern

6.3.1 *Unsuccessful application for a post at university*

Because Einstein graduated university with superior grades, he expected to obtain a post at university reasonably. The character of Einstein who halfway left the Luitpold Gymnasium disliking principle of education, was continued on his youth. At that time, it was common to call a Professor with honorific of "Herr Professor" but during period of attendance at the ETH, he did not call a Professor with honorific. He was considered as youth taking sassy attitude. Einstein's experimental plan on "the Earth's movement against the aether" was not permitted by Weber (Heinrich Friedrich Weber) who was the Professor taking charge of Einstein. One day, Weber said to Einstein "You are a very smart boy, but you have a serious defect of no hearing person's talk (Pais, 1982:44). The

Fig. 6.5.   Habicht, Solovine and Einstein (photograph in 1903).

term "smart" contains "clever" and also "cunning" in the meaning. Since occurrence like this, Einstein's enthusiasm and fascination for experiment gradually faded. Although the post of assistant was vacant, the Professor did not intend to offer the post to Einstein. He wrote to Onnes and Ostwald in Leiden and Leipzig, respectively letters applying for a post, but his applications were unsuccessful. Three other students in the same class as Einstein each immediately obtained position as assistant at the ETH.

In 1901, he got nationality of Swiss. On 19th May, he got a substitute teacher at industrial high school in Winterthur. Then, he was aware of possibility to keep effort and enthusiasm on science though he did not get a post at university, and gave up to get a post at university. He wrote to friend Grossmann a letter describing "I research on kinetic gas theory, and consider on relative motion of body against aether" (Pais, 1982:46). On 15th September, he got work of temporary position at private school in Schaffhausen.

In November of the year, for getting a doctoral degree, he submitted a thesis on kinetic gas theory to Zurich University. Then, ETH did not confer doctoral degree. However, his thesis was not permitted for doctoral thesis. This was his last failure.

6.3.2 *Grossmann's kindliness*

Because Grossmann said Einstein's difficulty in finding employment, to Grossmann's father, his father nominated Einstein to Haller (Friedrich Haller) President of the Patent Office in Bern. On 11th December 1901, the vacancy of one post in the Patent Office was announced officially. Immediately Haller interviewed Einstein, and informed to guarantee a post for Einstein.

In February 1902, Einstein resigned the public school in Schaffhausen, and relocated to Bern. Living expenses depended on the income of private teacher and few money sent from home. Students of private teacher, Solovine (Maurice Solovine) and Habicht (Konrad Habicht) became his close friends, and they regularly met to discuss on philosophy, physics and literature having simple meal. They called the meeting as "Akademie Olympia."

6.3.3 *Patent office*

On 16th June 1902, he served in the Patent Office. First he temporally employed as the third type of technical staff, and On 16th September 1904, he was employed as regular staff. On 1st April 1906, he became the second type of technical staff. The work there was to examine whether a patent applied was based on scientific principle or not.

In the Patent Office, during the daytime break, for preventing from wasting time, he avoided a contact to person, and was lost in thought on his research sparing time. Consequently, he was thought as a strange

Fig. 6.6. Einstein in 1904.

Fig. 6.7. Marcel Grossmann (1878–1936).

public official. No one did not think that afterward, he would accomplish great discovery.

After the Patent Office's working hours, in home he continued his theoretical researches. Because there was not academic space like university in the Patent Office, research at home was his reason for living. The process of accumulating little by little research work gave him sense of fulfillment. Though in 1901, he could not get a doctoral degree, he furthermore research and submitted thesis "Eine neue Bestimmung der Molekuldimensionen" to Zurich University in 1905 and got a doctoral degree. He dedicated the doctoral dissertation to Grossmann. Grossmann also got a doctoral degree at Zurich University.

6.3.4 *Mileva*

On the other hand, Einstein's private life was as follows. Before he relocated to Bern, he intended to marry with Mileva (Mileva Marity) with whom he discussed on science at ETH. She was born at Titel in South Hungary in 1875, and of Greek Catholic background. Einstein's mother neither at the time nor later liked her. Therefore, his parents strongly opposed to the marriage. By a brief and fatal heart disease his father fell. When he visited his father, his father finally consented to his son's marriage. His father passed away on 10th October 1902. On 6th January 1903, he married. On 14th May 1904, his eldest son Hans (Hans Albert) was born.

6.4 Publication of Three Papers in 1905

6.4.1 *Three papers in Annalen der Physik*

The research results which were accomplished at home, working at the Patent Office, were published as three papers in Annalen der Physik in Germany. The first paper was on the photoelectric effect entitled "Uber einen die Erzeugung und Verwandlung des Lichtes betreffenden heuristischen Gesichtspunkt," the second paper was on the special relativistic theory entitled "Zur Elektrodynamik bewegter Korper," and the third paper was on Brownian motion entitled "Uber die von der molekularkinetischen Theorie der Warme geforderte Bewegung von in ruhenden Flussigkeiten suspendierten Teilchen." All the papers were revolutionary improving previous physics.

6.4.2 *Elucidating photoelectric effect by energy quantum*

The first paper verified the concept of "energy quantum" proposed by Planck (Appendix 6.1). The concept of "energy quantum" insisted that "electromagnetic energy radiated by black body such as heat radiation, is integer times energy quantum, and discrete." The concept was incompatible with classical physics. According to electromagnetic theory by Maxwell and thermodynamics, electromagnetic energy was considered to consist of waves, and to be continuous and not to be discrete.

However, introducing the concept of "energy quantum," Einstein called light with energy quantum as "photon," and he was successful in theoretically elucidating the phenomena of "photoelectric effect" where an electron was emitted from metal irradiated by light. By this, the concept of "energy quantum" proposed by Planck became not hypothesis but truth. Instead of Newtonian mechanics which was indicated application limit in microscopic world such as atom, quantum mechanics based on the concept of energy quantum was developed at the beginning of the 20th century. In 1922, Einstein was awarded the Nobel prize for physics because of elucidating photoelectric effect. Photoelectric effect is applied to photo-sensors such as photoelectric tube and photo-electron multiplier (Shioyama, 2002).

Appendix 6.1 Energy Quantum

When Planck derived Planck's formula (Appendix 6.2) which theoretically solved heat radiation in 1900 that could not be explained by classical physics, he profoundly considered the reason why this formula inevitably held in this form, seeking the physical meaning of the basis. As a result, he was aware of the truth that energy could not take continuous value but take integer times inseparable unit of quantum. Energy quantum was related to frequency ν, and was given by hν. Where h is Planck constant. When n is nonnegative integer, energy is given by product of n by

Fig. 6.8. Max Karl Ernst Ludwig Planck (1858–1947) (photograph in 1890).

quantum. The concept of quantum at the beginning of 20th century led new quantum mechanics (Appendix 3.2).

Einstein's the second paper was on the special relativistic theory. This paper brought about a revolution in physics. The new theory will be described at Paragraph 6.6 in this chapter.

6.4.3 *Statistical property of Brownian motion*

Einstein's the third paper based on Boltzmann statistics proved existence of atoms. That is, Maxwell and Boltzmann founded "Kinetic gas theory" which assuming that gas consisted of many atoms or molecules, deriving physical states such as pressure of gas and thermal energy. Boltzmann extended the kinetic gas theory and developed "Boltzmann statistics." But there was left an inquiry of "does atom exist?" There was however only one phenomenon which estimated existence of atom. It was the phenomenon which was discovered by Brown (Robert Brown) observing particle of pollen. The phenomenon was Brownian motion where small particles suspended in a liquid were observed to jiggle around, and was published in 1828.

Einstein in the third paper considering "Brownian motion was brought based on the effect of millions of random collisions due to atom or molecules with all direction," theoretically derived trajectory of particle of pollen based on Boltzmann statistics, and indicated that the theoretical results coincided to measured results in Brownian motion. By this, existence of atom was verified.

Explanation 6.1 Brownian Motion

Brownian motion is the random motion of particles suspended in a medium (a liquid or a gas). In the one dimensional case, the probability density $\rho(x, t)$ of particle being in point x at time t, satisfies the diffusion equation:

$$\partial\rho(x, t)/\partial t = D[\partial^2\rho(x, t)/\partial x^2] \tag{1}$$

where D is the diffusion constant, given by

$$D = RT/(6\pi\mu a\, N_a) \tag{2}$$

where R is the gas constant, N_a Avogadro number, a the radius of a particle, μ is the viscosity of the medium, T absolute temperature.

The mean squared displacement from the equilibrium point x_0 is expressed by

$$\langle(x - x_0)^2\rangle = \int(x - x_0)^2\rho(x, t)dx \tag{3}$$

From (1) and (3), and using partial integral formula, we have

$$\partial\langle(x - x_0)^2\rangle/\partial t = \int(x - x_0)^2[\partial\rho(x, t)/\partial t]\, dx$$
$$= D\int(x - x_0)^2[\partial^2\rho(x, t)/\partial x^2]\, dx$$
$$= 2D \tag{4}$$

where we use $\int\rho(x, t)dx = 1$. From (4) we have

$$\langle(x - x_0)^2\rangle = 2Dt \tag{5}$$

Hence as Einstein indicated, the displacement of a Brownian particle is not proportional to the elapsed time, but rather to its square root.

Fig. 6.9. Brownian motion.

Appendix 6.2 Planck's Formula

Planck considered that black body radiation consisted of oscillators of electromagnetic vibration (mechanical system with sinusoidal vibration was called oscillator) (Tomonaga, 1952). Expressing frequency of oscillator of electromagnetic vibration by ν, the number $Z(\nu)$ of oscillators with frequency ν per unit volume was described as

$$Z(\nu) = (8\pi/c^3)\nu^2$$

by considering constraint for wave length of stationary wave in oscillator, where c is the velocity of light. Planck thought that energy E of an oscillator was non negative integer times energy quantum ε, having only discrete value. Based on his concept of discrete energy, and according to Boltzmann's principle, the average energy $\langle E \rangle$ was derived as

$$\langle E \rangle = \varepsilon/[\exp\{\varepsilon/(k_B T)\} - 1]$$

where T denotes absolute temperature of black body, and k_B denotes Boltzmann constant given by gas constant divided by Avogadro number.

Because energy quantum ε was expressed by $h\nu$, intensity of black body radiation $U(\nu)$ which was the product of number of oscillators $Z(\nu)$ by average of energy $\langle E \rangle$, was given by Planck formula as

$$U(\nu) = (8\pi h/c^3)\nu^3/[\exp\{h\nu/(k_B T)\} - 1].$$

Because numerator of $U(\nu)$ is proportional to the cube of ν, $U(\nu)$ increases with frequency. Because denominator of $U(\nu)$ exponentially

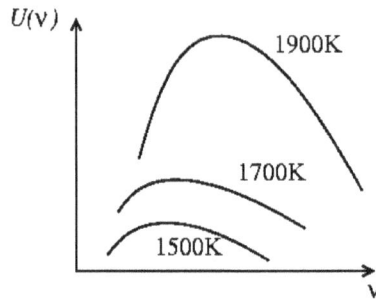

Fig. 6.10.   Relation between intensity of black body radiation U(ν) and frequency ν of electromagnetic wave.

increases with frequency and the increasing rate becomes superior to the increasing rate of numerator, $U(\nu)$ decreases at some frequency. Hence, $U(\nu)$ has a maximum value at some frequency. For black body with high temperature T, increasing rate of exponential function of denominator is small than low temperature T, the frequency at which $U(\nu)$ has maximum value shifts to higher side, according to Wien's displacement law. The formula coincided with experimental data of relation of intensity of black body radiation with frequency. Planck's success in deriving the formulation was due to introducing the concept of "energy quantum" in which energy could take only discrete value of integer times quantum.

6.5 Historical Background Before the Special Relativistic Theory

6.5.1 *Aether*

Before Einstein founded the special relativistic theory, he researched historical background of physics in detail. He summarized problem and contradiction left. Persons at the 19th century including Maxwell, considered that fundamental equations of electromagnetic theory held true only in inertial frames at rest against aether (Moller, 1959), (Yukawa & Tamura, 1955–1962:III). The inertial frame is rectangular coordinate system moving at constant velocity where Newton's first motion law "inertia's law" holds true. In the 19th century, it was considered that the aether penetrated every matters and vacuum, and was medium bearing every optical and electromagnetic phenomena (Hawking, 2001; Pais, 1982).

6.5.2 *Galilean transformation*

The inquiry whether aether existed or not, was left. In order to obtain the solution for the inquiry, the scientists performed optical experiments in inertial frame moving against aether. When inertial frame I' moves at velocity v in x direction against inertial frame I at rest in aether (Fig. 6.11), the velocity of light in x direction in inertial frame I' is smaller than that in inertial frame I. That is, velocity of light is different in different inertial frame, expecting to yield the displacement of interference fringe due to phase difference of two lights one of which is in direction of v and the other of which is perpendicular to v (Appendix 6.3). Using Galilean trans-formation, optical experiments inspecting whether the expected displace-ment of interference fringe occurred or not, were performed. Galilean transformation is given by

$$x' = x - vt, y' = y, z' = z, t' = t.$$

6.5.3 *Optical experiment by Michelson and Morley*

In 1859, experiment of Armand Hippolyte Louis Fizeau and in 1865, experiment of Jean Bernard Leon Foucault were performed. When

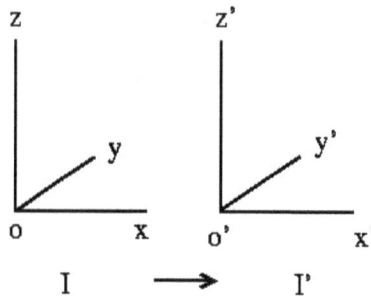

Fig. 6.11. Inertial frame I' moving at constant velocity v in x direction against inertial frame I.

velocity of light is expressed as c, the method measuring the square of v/c was tried by Albert Abraham Michelson for the first time in 1881. In 1887, 6 years later, the method was furthermore improved by Michelson and Edward William Morley (Appendix 6.3). Their improvement is as follows: In the first improvement, they set equipment on grand stone (1.5 m on every side, 0.3 m thickness) which floated in stratum of mercury, so as to rotate all equipment without distortion and to be not influenced by vibration. In the second improvement, they set many reflective mirrors by which light was reflected repeatedly, so as to obtain 10 times optical path, because estimated displacement of interference fringe in inertial frame moving against aether was so small, and the displacement of interference fringe was proportional to optical path (Appendix 6.3). Furthermore, they covered all the optical equipment with wood cover for preventing from air stream and temperature change. Thus, they fabricated the most precise experimental equipment by the devised grand improvement.

In their experiment, the displacement of interference fringe was so smaller than the estimated one (Appendix 6.3). The displacement of interference fringe estimated on assuming that the velocity of light was different in different inertial frame, did not occur. Logically, when proposition "if A occurs, then B occurs" is true, then the contraposition "if B does not occur, then A does not occur" is also true. In this case, A is "velocity of

Fig. 6.12.   Armand Hippolyte Louis Fizau (1819–1896).

Fig. 6.13. Jean Bernard Leon Foucault (1819–1868).

light being different in different inertial frames" and *B* is "displacement of interference fringe." Because the contraposition is true, it is concluded that "if there is no displacement of interference fringe, then velocity of light is same in different inertial frames." The experimental results supported "relativity principle" which insists that velocity of light should be same in all inertial frames.

Appendix 6.3 Michelson–Morley Experiment

As figure half mirror HM was set at 45 degree angle in light path from light source S (Yukawa & Tamura, 1955–1962:III). By the HM, light was split and each light proceeded with mutually 90 degree angle. Each light reflected at *M*1 and *M*2 respectively, went back to half mirror, and then proceeded in same direction to detector *D*. Because mirror *M*1 and *M*2 were not precisely perpendicular to light path, interference fringe was observed corresponding to phase difference of two lights at two dimensional space in detecting plane. If measuring system moved at velocity v in direction from HM to *M*2 and velocity of light was different in different inertial frame, there occurred difference Δt in time for proceeding optical paths of each light, and the displacement of interference fringe should be observed corresponding to the difference in time, where optical path from half mirror to each reflective mirror was same d. Define c as velocity of light, β as v/c, ν as frequency and T as period of light wave. Then time difference Δt was given by $d\beta^2/c$ (Moller, 1959:25), and phase difference ΔF is given by $\nu\Delta t$ which is written by $\Delta t/T$. The ratio of displacement of interference fringe to interval of neighboring fringes was given by $\Delta t/T$. Define λ as wave length of light. Because velocity of light is proceeding distance per second, $c = \lambda\nu$. Hence ΔF is given by $d\beta^2/\lambda$. If the interval of neighboring fringes was used as a unit, mentioned below, $\Delta F = \Delta t/T$ meant the displacement of interference fringe. When all the equipment were rotated 90 degree, the phase difference became — ΔF, which indicating the corresponding displacement of interference fringe to be $2\Delta F = 2d\beta^2/\lambda$. In Michelson–Morley experiment, v was the velocity of the Earth 3×10^4 m/s, d was 11 m, and λ was 5.89×10^{-7} m. Then the displacement of interference fringe was estimated to be 0.37. But the displacement of interference fringe measured in the experiment was about 1/40 very smaller than the estimated value (Yukawa & Tamura, 1955–62:III). In Fig. 6.14, though reflective mirrors $M1$ and $M2$ are represented, as mentioned above, in Michelson–Morley experiment, many reflective mirrors were set to lengthen optical path d by repeating reflection.

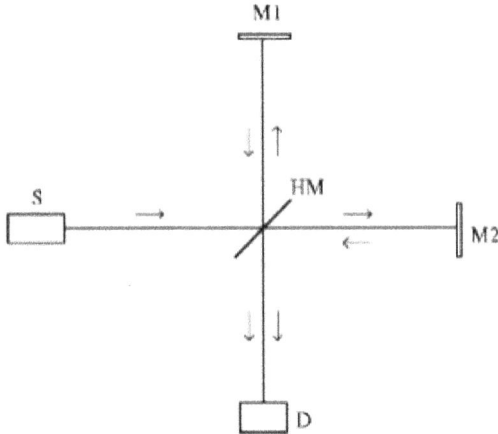

Fig. 6.14. Michelson–Morley experiment. S: light source, D: detector, HM: half mirror, $M1$, $M2$: reflective mirror.

Fig. 6.15. Albert Abraham Michelson (1852–1931).

Fig. 6.16. Edward William Morley (1838–1923).

Appendix 6.4 Details Until Formulation of the Special Relativistic Theory

"Relativity principle" means that all physical laws hold true in any inertial frame. Concerning elementary laws of electromagnetic theory, "relativity principle" was not considered to hold true under Galilean transformation. The reason is as follows. Maxwell's equations contain light velocity with constant c (equal to velocity of electromagnetic wave because light is electromagnetic wave). Hence by relativity principle, if Maxwell's equations hold true in any inertial frame, light velocity should be constant c irrespectively of light source motion. This contradicts the usual idea on motion. For example, if inertial frame I' moves in direction of light against inertial frame I, then light velocity in I' is smaller than that in I.

Consequently, if relativity principle is recognized, then previous idea about transformation between time and space in two inertial frames mutually moving, should be improved. Before the improvement was performed, necessity of the improvement should be verified, This verification should be obtained only by experimental results, and optical experiment was the most suitable for the objective as mentioned above.

If relativity principle is recognized, then aether called as coordinate system at absolute rest, loses physical meaning, because when relativity principle is recognized, equivalence of all inertial frames is requested. Furthermore, Galilean transformation is denied. This changes the base of natural description. Einstein improved the previous idea on time-space, and formulated new theory recognizing relativity principle.

6.6 The Special Relativistic Theory

6.6.1 *Relativity principle*

The details till the formulation of the special relativistic theory under the historical background is described in Appendix 6.4. "Relativity principle" is defined as that all laws of physics hold true in any inertial frame. In 1905 Einstein published elementary paper on the special relativistic theory where he formulated for the first time new theory recognizing "relativity principle," and derived many results. He recognized relativity principle defined as that natural laws appear in same form in infinitely many inertial frames mutually moving linearly at constant velocity. Relativity principle insists the equivalence of all inertial frames, and it demands the equivalence not only in mechanical phenomena such as Newtonian mechanics but also in electromagnetic phenomena such as electromagnetic theory of Maxwell.

6.6.2 *Invariant light velocity*

Based on relativity principle, Maxwell's equations should hold true in same form in any inertial frame. Considering these equations containing constant light velocity, light velocity should be constant in vacuum irrespectively of light source moving. Then because light velocity is constant in any inertial frame, light velocity is the invariant for coordinate transformation between inertial frames. Hence it was requested to determine the coordinates of time-space of inertial frame so that light velocity became invariant for coordinate transformation between inertial frames. For satisfying the request, Galilean transformation became inappropriate. Thus Einstein derived the new concept on time-space (Yukawa & Tamura, 1955–62:III).

6.6.3 *Lorentz transformation*

The conclusion which was estimated assuming the existence of aether called as "coordinate system in absolute rest," were denied by experimental results. Hence for the purpose of verifying relativity principle, Lorentz considered as follows. Considering what assumption should be introduced so that conclusions which were estimated assuming the existence of aether, suited with relativity principle, for the first time he derived formula called as Lorentz transformation in 1904.

In the case of Fig. 6.11 with velocity v, it is given by

Lorentz transformation: $\gamma \equiv 1/(1 - v^2/c^2)^{1/2}$

$$x' = (x - vt)\gamma, \; y' = y, \; z' = z, \qquad x = (x' + vt')\,\gamma$$
$$t' = (t - vx/c^2)\,\gamma, \qquad t = (t' + vx'/c^2)\,\gamma$$

Proof. From the uniformity of time-space, the coordinates x', y', z', t' are transformed to the linear combination of x, y, z, t. Since $x' = 0$ for $x = vt$, $y' = 0$ for $y = 0$, $z' = 0$ for $z = 0$, we have

$$x' = A(x - vt), \; y' = By, \; z' = Cz, \; t' = Dx + Ey + Fz + Gt \qquad (6.1)$$

Considering the light starting O, O' of origin in I and I'-frame at $t = t' = 0$, we have

$$x^2 + y^2 + z^2 - c^2t^2 = 0, \ x'^2 + y'^2 + z'^2 - c^2t'^2 = 0 \qquad (6.2)$$

From (6.1) and (6.2), we have $k(v)^2 \equiv A^2 - c^2D^2 = B^2 = C^2 = G^2 - A^2v^2/c^2$, $c^2DG = -vA^2$, $E = F = 0$. Hence we have

$$x' = \gamma k(v)(x - vt), \ y' = k(v)y, \ z' = k(v)z, \ t' = \gamma k(v)(t - vx/c^2) \qquad (6.3)$$

$$\gamma \equiv 1/(1 - v^2/c^2)^{1/2}$$

Considering inverse x-axis, x'-axis, z-axis and z'-axis, then $x \to -x$, $z \to -z$, $x' \to -x'$ and $z' \to -z'$. Hence $z' = k(v)z$. However, in this case, I-frame move with velocity v against I'-frame, therefore $z = k(v)z'$. Hence we have $z' = k(v)^2z'$; $k(v) = 1$ and get the Lorentz transformation.   □

Einstein independently derived Lorentz transformation instead of Galilean transformation so that constant light velocity held true in all inertial frames. Lorentz transformation was for the first time derived by Lorentz, but Einstein derived the transformation based on fundamental law of the special relativistic theory, and indicated the new physical meaning.

Fig. 6.17.   Hendrik Antoon Lorentz (1853–1928).

6.7 Consequence of the Special Relativistic Theory

6.7.1 *Contraction of moving body*

From equations of Lorentz transformation, "contraction of moving body" was derived. The square root of $[1 - (v/c)^2]$ was called as Lorentz contraction, where v is body's velocity against the inertial frame I and c is light velocity. The length of moving direction was given by the length at rest multiplied by Lorentz contraction (that is, contracted in direction of velocity). Observer staying in static I-frame observes the moving body's contraction, with I-frame's measure.

Proof of contraction. We consider a stick positioned along x'-axis in I'-frame. Let the position of its edges be x_1', x_2'. Let the length be l_0: $x_2' - x_1' = l_0$. From Lorentz transformation,

$$x_2' - x_1' = (x_2 - vt)\gamma - (x_1 - vt)\gamma = (x_2 - x_1)\ \gamma$$

$$\gamma = 1/(1 - v^2/c^2)^{1/2}$$

$$x_2 - x_1 = l_0(1 - v^2/c^2)^{1/2} \qquad \square$$

6.7.2 *Delay of moving clock*

Furthermore, from Lorentz transformation, "delay of moving clock" was derived. Advance of moving clock $\Delta\tau$ with velocity v was given by advance of clock at rest Δt multiplied by Lorentz contraction. That is, advance of moving clock $\Delta\tau$ is delayed compared to advance of clock at rest Δt. For example, moving radioactive substance emits radiation longer

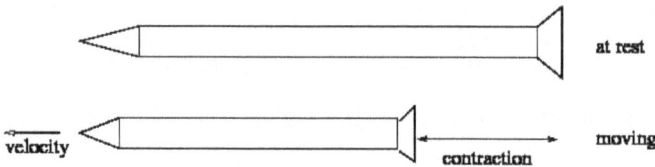

Fig. 6.18   Contraction of moving body.

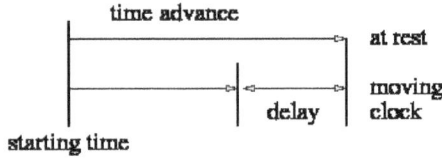

Fig. 6.19. Delay of moving clock.

time than the case at rest because advance of clock is delayed than in case at rest, and the life time is lengthened (Moller, 1959:45).

$$\Delta\tau = \Delta t[1 - (v/c)^2]^{1/2}$$

6.7.3 *Relativistic mass of moving body*

Maxwell's equations which are elemental equations in electromagnetic phenomena, are invariant to Lorentz transformation, harmonizing with relativity principle. But it was necessary to add change in elemental equations of Newtonian mechanics for the purpose of harmonizing the equations of Newtonian mechanics with relativity principle. From this change, relativistic mass m of moving body of static mass m_0 with velocity u, was derived from momentum conservation law as

$$m = m_0/[1 - (u/c)^2]^{1/2}: \text{ relativistic mass}$$

Proof. We consider elastic collision between two particles with velocities in direction of x'-axis, u_1' and $u_2' = -u_1'$ in I'-frame. Let velocities of these particles in I-frame be u_1 and u_2. Lorentz transformation for u in I-frame and u' in I'-frame;

$$u' = (u - v)/(1 - uv/c^2), \ u = (u' + v)/(1 + u'v/c^2) \tag{6.4}$$

The momentum conservation law is described as

$$m(u_1)u_1 + m(u_2)u_2 = [m(u_1) + m(u_2)]v \tag{6.5}$$

where $m(u)$ denotes mass of particle with velocity u and immediately after collision both particles are static in I'-frame. From (6.4) and (6.5) we have

$$m(u_1)/m(u_2) = (1 - u_2 v/c^2)/ (1 - u_1 v/c^2) \qquad (6.6)$$

We obtain from Lorentz transformation the following:

$$(1 - u'^2/c^2)^{1/2} = (1 - v^2/c^2)^{1/2}(1 - u^2/c^2)^{1/2}/(1 - uv/c^2)^{1/2} \qquad (6.7)$$

From (6.6) and (6.7), we have

$$m(u_1)/m(u_2) = (1 - u_2^2/c^2)^{1/2}/(1 - u_1^2/c^2)^{1/2}$$

In case of $u_2 \ll c$, $m(u_2)$ becomes the static mass m_0 in Newtonian mechanics. Hence

$$m(u_1) = m_0/(1 - u_1^2/c^2)^{1/2} \qquad \square$$

6.7.4 *Mass-energy equivalence*

From the relation in Lorentz transformation about momentum p and energy E, Einstein indicated that relativistic mass was given by E/c^2, and indicated "mass-energy equivalence." Hence, mass point with mass m had energy E in the following.

$$E = mc^2: \text{mass-energy equivalence}$$

Proof. Lorentz transformation in case where the velocity v of inertial frame I' against inertial frame I is in other direction than x-axis, is expressed by

$$x' = x + v[\{(x \cdot v)/v^2\}\{(1 - v^2/c^2)^{-1/2} - 1\} - t(1 - v^2/c^2)^{-1/2}] \qquad (6.8\text{-}1)$$

$$t' = (1 - v^2/c^2)^{-1/2}\{t - (v \cdot x)/c^2\} \qquad (6.8\text{-}2)$$

From (6.8-1) and (6.8-2), Lorentz transformation for the velocity of particle $u = dx/dt$ is derived as:

$$u = [u' + v[\{(u' \cdot v)/v^2\}\{(1 - v^2/c^2)^{-1/2} - 1\} + (1 - v^2/c^2)^{-1/2}]]$$
$$/[(1 - v^2/c^2)^{-1/2}\{1 + (v \cdot u')/c^2\}].$$

We have the Lorentz transformation for momentum $p = m_0 u/(1 - u^2/c^2)^{1/2}$ and energy $E = m_0 c^2/(1 - u^2/c^2)^{1/2}$ of free particle (Moller, 1959: 69):

$$p = p' + (v/v^2)[(v \cdot p')\{1 - (1 - v^2/c^2)^{1/2}\} + E'v^2/c^2]/(1 - v^2/c^2)^{1/2} \qquad (6.9)$$
$$E = [E' + (v \cdot p')]/ (1 - v^2/c^2)^{1/2} \qquad (6.10)$$

When we set I'-frame so that $p' = 0$; $(u' = 0)$ and particle's relative velocity against I-frame is v, we have

$$p = E_0 \, v/[c^2(1 - v^2/c^2)^{1/2}], \; E = E_0/(1 - v^2/c^2)^{1/2} \qquad (6.11)$$

The ratio of p to v is recognized as mass m which is from (6.11) given by

$$m = E/c^2; \; E = mc^2 \qquad \qquad \square$$

The mass-energy equivalence has realistic meaning only when extinction process of particle exists. In the process, energy equivalent to decrease of mass of particle is changed to other particle's kinetic energy. If extinction process exists in nature, then on extinction of mass, energy is released (Møller, 1959) (Explanation 6.2). If extinction process of particle occurs in chain reaction, then enormous amount of energy is released. Afterward, from horror to Nazi, in 1939 Einstein suggested to President Roosevelt (Franklin D. Roosevelt) possibility of production of atomic bomb on the basis of this theoretical conclusion.

6.7.5 *Mass defect and nuclear fission*

"Mass-energy equivalence" was experimentally verified in nuclear fission reaction where nucleus changed to other nucleus with different mass

Fig. 6.20.   John Douglas Cockcroft (1897–1967).

Fig. 6.21.   Ernest Thomas Sinton Walton (1903–1995).

defect (Explanation 6.3). The verification of mass-energy equivalence was performed in the experiment by Cockcroft (John Douglas Cockcroft) and Walton (Sinton Walton) at Cavendish Laboratory.

At experiment, on lithium whose mass is 7.0166, being impacted with proton whose mass is 1.0076, with high velocity, proton plunged into lithium nuclear, and compound nucleus was formed. New nucleus formed by adding incident particle to nucleus, whose interior energy is higher with added energy, is called as compound nucleus. In their experiment, because of compound nucleus being unstable, compound nucleus divided into two α particles (Helium nucleus) whose mass is 2×4.0028, with high velocity. By measuring the lost mass which is 0.0186 (= 7.0166 + 1.0076 − 2 × 4.0028) corresponding to 27.7 × 10^{-6}erg (Explanation 6.2) and quantity (27.6 ± 0.05) × 10^{-6}erg which was obtained by subtracting incident proton's kinetic energy from α particle's kinetic energy obtained after nuclear fission reaction, coincidence of the energy equivalent to lost mass with kinetic energy, was verified. Thus, mass-energy equivalence was verified (Mϕller, 1959).

Explanation 6.2 Nuclear Fission

Nuclear fission is a reaction in which the nucleus of an atom splits into two or more smaller nuclei.

For example, U^{235} absorbing neutron n is unstable and then splits into yttrium Y^{95} and iodide I^{139} together with heigh speed neutron:

$$U^{235} + n \rightarrow Y^{95} + I^{139} + 2n$$

The product materials in nuclear reaction of U^{235} vary in distribution which has two peaks with mass 95 and 139.

At this reaction, the energy $E = \Delta mc^2$ corresponding to the lost mass Δm in nuclear reaction is released (Yukawa & Tamura, 1955–62:III280).

The mass of U^{235} which can produce average one neutron in nuclear fission, is called as critical mass. The U^{235} with above critical mass can produce chain reaction.

Explanation 6.3 Mass defect

The difference between the real mass of nucleus and mass of sum of proton and neutron which constitute nucleus, is called as mass defect (Fig. 6.22). Necessary energy to break up nucleus into particles is called as binding energy. When binding energy is expressed by ΔE, mass defect is given by $\Delta E/c^2$. Mass of sum of particles consisting nucleus being broken up is not equal to original nucleus's mass, and the difference of mass corresponds to binding energy (Mϕller, 1959). The average binding

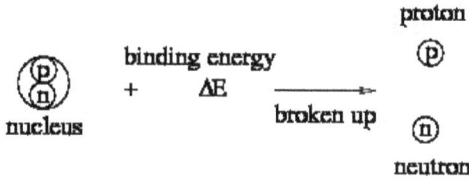

Fig. 6.22. Binding energy. ΔE: binding energy; p: proton; n: neutron. The nucleus is broken up by ΔE.

energy per nucleus is about 8 Mev for ordinary nucleus (Yukawa & Tamura, 1955–62:III280).

6.8 Research at University

6.8.1 *Getting a post at university*

In 1907, Einstein solved problem in specific heat of solid. In 1908, working at the Patent Office, he got a post of Private lecturer (Privatdozent) at Bern University. Taking the post meant that he had only right of teaching without belonging to the department of university. Salary was not paid from university, and salary was few fee paid by attendants. But this post was the first post in academic place for him. Then his sister studying in Bern University attended his lecture. On 21st December 1908, she submitted thesis on Romance languages to Bern University, and got PhD degree in Romance languages magna cum laude.

On 15th October 1909, he got a post of Associate Professor of theoretical Physics at Zurich University. He resigned Bern University and the Patent Office. Then he was recognized as leading scientist. He was invited as Professor of the Karl-Ferdinand University in Prague. In March 1911, he and family arrived at Prague, and was inaugurated from 1st April.

In 1907, Grossmann became a full Professor of geometry at ETH. In 1911, he became dean of the mathematics-physics section of the ETH. The young dean's first work was to sound Einstein out "his thought in returning to Zurich, this time to the ETH" (Pais, 1982:208–209). Einstein immediately informed his thought to teach at ETH.

6.8.2 *Professor at ETH*

In 1912, he was invited as Professor at ETH, and he relocated to Zurich from Prague. Before relocating to Zurich, the University of Utrecht invited him, but he rejected it. Though on graduating ETH in 1900 he was not successful in application for a post at university, 12 years after the graduation he was inaugurated as Professor at ETH.

6.8.3 *Founding the general relativistic theory with Grossmann*

From this year, with Grossmann, he set about founding the general relativistic theory which generalized the special relativistic theory. Using tensor calculus and Riemannian geometry which was the theory about curved space and surface, he developed gravity theory. In 1913, he published a paper on the general relativistic theory "Draft of the general relativistic theory and gravity theory (Entwurf einer verallgemeinerten Relativitatstheorie und einer Theorie der Gravitation)" with Grossmann. In the paper, they proposed an idea that gravity was related to distortion of time-space. At that time, they did not yet find the gravitational field equation which related distortion of time-space to matter distribution causing gravity. As mentioned below, in 1915 Einstein found it.

Explanation 6.4 Tensor Analysis

Tensor analysis expands vector analysis to tensor field such as time-space. It had been developed by Gregorio Ricci-Curbastro and his pupil Tullio Levi-Civita. Einstein adopted it in developing the general relativistic theory, using Ricci curvature tensor R_{ik}, scalar curvature R and covariant differentials (Paragraph 6.9).

6.8.4 *Berlin*

In April 1914, because Einstein was invited as Professor of the University of Berlin with nomination by Planck, he relocated to Berlin from Zurich.

Immediately afterward, his wife and two sons (Hans Albert and Eduard) came to him. But his wife and two sons went back to Zurich, and he started living apart from family. Due to living apart, he devoted himself to theoretical research. During 18 years after then, he stayed in Berlin. The Kaiser Wilhelm Institute for physics inaugurated him as director. Though he came back to Germany, he did not intend to get citizenship of Germany in situation having citizenship of Swiss gotten in 1901. He wrote in letter to Lorentz "A post in Berlin frees me of all obligations so that I can devote myself freely to thinking." He wrote in letter to friend Zannger (Heinrich Zannger) director of the Institute for Forensic Medicine at the University of Zurich, "Contact with the colleagues in Berlin might be stimulating, and especially the astronomers are important for me," indicating that Einstein then was interested in the bending of light (Pais, 1982:240).

On 2nd July 1914, at the Prussian Academy, he delivered inauguration lecture for Professor at Berlin University. He admired Planck as "Planck's idea of energy quantum introduced improvement in microscopic field," and then delivered lecture on his own relativistic theory. On reply to his lecture, Planck finished his speech with "experimental results on "the bending of light due to gravity" theoretically predicted by Einstein is expected to be obtained by the total solar eclipse on 21st August" (Pais, 1982:242). The first World War on 1 August out broke the expectation. Though at that time the prediction of the bending of light due to

Fig. 6.23. David Hilbert (1862–1943).

gravitation was not yet correct, the prediction was made correctly in the complete version of the general relativistic theory in 1915, being verified by celestial observation in 1919 (Paragraph 6.11).

6.8.5 *Hilbert*

On 25th November 1915, Einstein submitted the complete version of the general relativistic theory to Annalen der Physik. However 5 days before his submission, Hilbert (David Hilbert) submitted a paper which contained gravitational field equations of the general relativistic theory, and there might arise a trouble who had priority. In fact, on visiting Gottingen, Einstein discussed with Hilbert on his idea. Afterward Hilbert found gravitational field equations several days before Einstein found it. Nevertheless, Einstein's priority is recognized now because Einstein related gravity to distortion of time-space (Hawking, 2001:30). In 1916, the submitted paper above mentioned "Basis of the general relativistic theory (Die Grundlage der allgemeinen Relativitatstheorie)" which was the first and the most important paper, was published.

6.9 The General Relativistic Theory

6.9.1 *General relativity principle*

"Special relativity principle" means equivalence of all inertial frames moving at mutually constant velocity and without distortion. On the other hand, "general relativity principle" means equivalence of not only inertial frames but also accelerative frames. Accelerative frames are the systems which move with acceleration against inertial frame. For example, the system rotating against inertial frame and with centrifugal force is an accelerative frame.

In an accelerative frame rotating against inertial frames, apparent gravity called non-permanent gravity of centrifugal force occurs. Einstein introduced "equivalence principle" which insists equivalence between non-permanent gravity and permanent gravity such as gravity due to the Sun. In order to accept general relativity principle, it is necessary to use general curve coordinates instead of rectangular coordinates, taking into account the influence of acceleration whether due to permanent or non-permanent gravity (Moller, 1959:223).

On the basis of concept accepting general relativity principle, Einstein founded the general relativistic theory generalizing the special relativistic theory.

6.9.2 *Minkowski 4 dimensional world*

To describe natural phenomena, Einstein used Minkowski four dimensional world where three-dimensional spatial coordinates (X, Y, Z) and temporal coordinate (T) in inertial frames without distortion were unified to X_i (i = 1, 2, 3, 4) = $(X, Y, Z, c\,T,\ c$: light velocity). Minkowski four dimensional world was called as "World" (Explanation 6.5). The square of segment (four dimensional distance between two neighboring points) in Minkowski four dimensional world without distortion was given by subtraction of the square of segment dX_4 (= cdT) on temporal coordinate X_4 (= cT) from the sum of three squares of segment dX_i (i = 1, 2, 3) (one-dimensional distance between two neighboring points). On the other hand, in accelerative frames, the square of the segment $d\sigma^2$ in Minkowski four dimensional world expressed by general coordinate x^i (i = 1, 2, 3, 4) with distortion, was given by the product of two among segment dx^i (i = 1, 2, 3, 4) of general curve coordinate, multiplied coefficient g_{ij} that is,

$$d\sigma^2 = g_{ij}dx^i dx^j \ (i, j = 1, 2, 3, 4)$$

which was total summation for all pair (i, j). The coefficient g_{ij} $(i, j = 1, 2, 3, 4)$ was called as metric tensor. The general curve coordinate x^i (i = 1, 2, 3, 4) is function of X_i (i = 1, 2, 3, 4). In Minkowski four dimensional world without distortion, it holds true that $x^i = X_i$ (i = 1, 2, 3, 4) and $g_{11} = g_{22} = g_{33} = 1$, $g_{44} = -1$, g_{ij} with different i and j is zero. Metric tensor plays important role in the general relativistic theory as mentioned below.

The gravitational equation which decides the relation between the geometry of time-space and matter distribution, was found by Einstein in 1915 as mentioned above (Paragraph 6.8 in this chapter). The gravitational field equation (Explanation 6.6) contains term concerning distortion of time-space due to gravity and term of mass-energy. In the case of weak gravity field and static mass distribution, the gravitational field equation becomes Newton's gravitational theory which is expressed by term of Laplacian of gravitational potential and term of mass distribution.

Metric tensor determines geometry (generally, non-Euclid geometry) of Minkowski four-dimensional world in accelerative frame. It was indicated that metric tensor depended on non-permanent gravity such as centrifugal force and permanent gravity. That is, gravity influences metric tensor, and causes distortion of time-space.

In the world where Minkowski four-dimensional world is not flat and time-space distorts due to gravity, the path giving minimum distance between two points is not linear line but curve. In the world, Euclid geometry does not hold true, and the sum of interior angles of triangle is smaller than 180 degree (π radian). When accelerative frame due to non-permanent gravity transfers to inertial frame, non-permanent gravity extinguishes.

Explanation 6.5 Minkowski four-Dimensional World

Fusion of spatial coordinate (*X, Y, Z*) and temporal coordinate *T* is called as Minkowski four dimensional world. Here, it is assumed that there is no distortion of time-space. Four coordinate axis are mutually perpendicular. In Fig. 6.24 for convenience, three-dimensional spatial coordinate axis (*X, Y, Z*) are expressed by a plane perpendicular to time axis-*T*. The trajectory of a body at rest is parallel to time coordinate axis in Minkowski four-dimensional world. The trajectory of event travelling in future direction at light velocity is upper cone in Fig. 6.24. The trajectory of event travelling from the past is lower cone. This cone is called as light cone.

Because velocity of body is less than light velocity, world line which is trajectory of moving body in Minkowski four-dimensional world, exists inside of light cone.

6.9.3 *World line and geodesic line*

A point of Minkowski four-dimensional world, that is, a point representing "when and where" is called as "world point." The moving path of particle in Minkowski four-dimensional world is called as "world line."

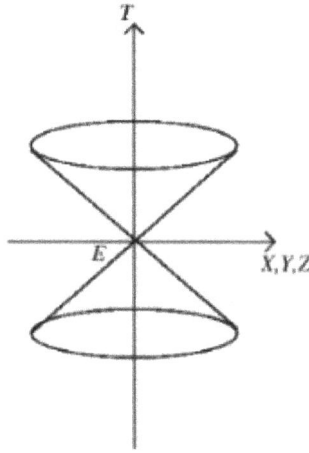

Fig. 6.24. Light cone.

The world line where distance between two world points in Minkowski four-dimensional world is minimum value, is called as "geodesic line." For the purpose of generalizing linear line giving minimum distance line in the world where there is no distortion of time-space as inertial frame, in the general relativistic theory, geodesic line is defined as minimum distance line. It was indicated that world line of particle free falling only under gravity was geodesic line. The geodesic line was given as solution of Euler equation derived from variational principle. Though in Newtonian mechanics "inertia's law" is motion law of body without force, in the general relativistic theory, "inertia's law" is replaced by "a body only under gravity moves along geodesic line."

6.9.4 *Relation between* g_{k4}, *k = 1, 2, 3, 4 and* χ, γ_k, *k = 1, 2, 3*

Gravitational acceleration is derived from gravitational potential. As gravitational potential, there are gravitational scalar potential χ and gravitational vector potential γ_k, $k = 1, 2, 3$. From Euler equation determining geodesic line, gravitational acceleration is expressed with metric tensor g_{44} and g_{k4}, $k = 1, 2, 3$, indicating correspondences as (Moller, 1959:243):

$$g_{44} = -(1 + 2\chi/c^2), \quad \gamma_k = g_{k4}/(-g_{44})^{1/2}, \, k = 1, 2, 3,$$

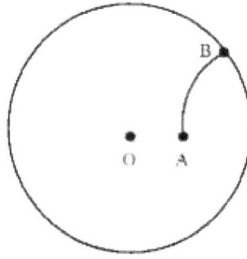

Fig. 6.25.   Geodesic line connecting points A and B in accelerative frame rotating.

where though χ is expressed by (χ + constant) giving the same gravitational acceleration −grad χ, the potential χ is normalized so that the constant is determined for g_{44} to be −1 which is the value in no gravity.

Although gravitational acceleration α is expressed using χ and γ_k, $k =$ 1, 2, 3, when spatial coordinate is perpendicular to time coordinate axis, or gravitational vector potential does not depend on time, α is given by the spatial gradient of χ multiplied by −1; $\alpha = -$grad χ. For example, in accelerative frames rotating with rotation angular velocity ω, χ of particle located at radius r, is given by $-(r\omega)^2/2$. Hence, α is $r\omega^2$, and its direction is direction of increasing r. That is, it is equal to acceleration of centrifugal force. Metric tensor appearing in segment of geodesic line is influenced by gravity, and distortion of time-space occurs, geodesic line being not linear line but curve as Fig. 6.25.

Explanation 6.6 Gravitational Field Equation

In the general relativistic theory, the gravitational field equations relate the geometry of space time to the distribution of matter within it. The gravitational field equation is expressed in the form:

$$M_{ik} = -\kappa T_{ik}, \ i, \ k = 1, 2, 3, 4 \tag{1}$$

where T_{ik} is the energy-momentum tensor, κ is the Einstein gravitational constant defined as

$$\kappa = 8\pi k/c^4 \tag{2}$$

where k is the Newtonian constant of gravitation and c is the velocity of light in vacuum.

In Newtonian gravitational field equation, the following Poisson equation holds:

$$\Delta\chi = 4\pi k\mu, \ \mu: \text{ mass causing gravitation.}$$

Since χ relates to g_{44}, M_{ik} is considered to have the form which contains the second derivative of metric tensor. Hence, we have the following relation:

$$M_{ik} = R_{ik} + c_1 R g_{ik} + c_2 g_{ik} \tag{3}$$

where R_{ik} is Ricci curvature tensor and R is the scalar curvature. The tensor is symmetric second-degree tensor that depends only on the metric tensor and its first and second derivatives.

The left hand side of equation (1) is the geometric quantities expressing the stress of space time, the right hand side expresses the distribution of matter which causes gravitation.

The coefficient c_1 is determined from energy-momentum conservation law to be $-1/2$, and c_2 is set to be $-\lambda$ invariant constant: cosmological constant (Moller, 1959:278, 283, 308). Hence we have

$$R_{ik} - (1/2)R g_{ik} - \lambda g_{ik} = -\kappa T_{ik} \tag{4}$$

Using

$$R_i^k = g^{kn} R_{in}, \ T_i^k = g^{kn} T_{in}, \ g_i^k = g^{kn} g_{in} = \delta_i^k \tag{5}$$

where indicator in two times appearance meaning summing up with respect to the indicator, for example,

$$g^{kn} R_{in} = g^{k1} R_{i1} + g^{k2} R_{i2} + g^{k3} R_{i3} + g^{k4} R_{i4},$$

$$g^{ik} = A_{ik}/g = g^{ki} \tag{6}$$

$$A_{ik}: \text{adjugate matrix determinant} \tag{7}$$

$$g: \text{determinant of matrix } (g_{ik}) \tag{8}$$

$$\delta_i^k = 1 \; i = k$$
$$= 0 \; i \neq k \tag{9}$$

(4) is written as

$$R_i^k - (1/2)R\delta_i^k - \lambda\delta_i^k = -\kappa T_i^k \tag{10}$$

By contracting (10), we have

$$R + 4\lambda = \kappa T \tag{11}$$

because $R_i^i = R$, $T_i^i = T$, $\delta_i^i = 4$. From (4) and (11), we have

$$R_{ik} + \lambda g_{ik} = -\kappa[T_{ik} - (1/2)\, Tg_{ik}] \tag{12}$$

In the case of spherically symmetric and static field, the solution of (12) is given by Schwarzschild solution (Moller, 1959:319) (Explanation 6.8).
The gravitational scalar potential χ influences metric tensor as:

$$g_{44} = -(1 + 2\chi/c^2) \tag{13}$$

Weak field

In the case of weak field, metric tensor is written by

$$g_{ik} = H_{ik} + h_{ik} \tag{14}$$

where

$$H_{ik} = 0 \; i \neq k$$
$$= 1 \; i = k = 1, 2, 3$$
$$= -1 \; i = k = 4 \tag{15}$$

h_{ik} and its derivative being small, its square negligible. In case of static mass distribution (Møller, 1959:309),

$$R_{44} = -\Delta\chi/c^2, \qquad T = -\mu c^2, \qquad \Delta\text{: Laplacian,}$$

$$T_{44} - (1/2)H_{44}T = (1/2)\mu c^2, \qquad T_{44} = \mu c^2 \tag{16}$$

From (12) and (16), we have by $i = k = 4$

$$\Delta\chi - \lambda c^2 \, g_{44} = (\kappa c^4/2)\mu \tag{17}$$

Neglecting the second term of left hand side because of small λ, (17) becomes a Newtonian Poisson equation. Hence, the limit of weak gravitational field yields Newtonian Poisson equation with μ: mass of matter,

$$\Delta\chi = (\kappa c^4/2)\mu = 4\pi k\mu \tag{18}$$

Using $\Delta = \text{div grad}$, and $\mathbf{F} \equiv -\mathbf{grad}\chi$, we have

$$\text{div } \mathbf{F} = -(\kappa c^4/2)\mu = -4\pi k\mu \tag{19}$$

By Gauss theorem, absolute value of $\mathbf{F}$ is given by

$$|\mathbf{F}| = \kappa c^4\mu/(8\pi r^2) = k\mu/r^2 \tag{20}$$

Hence, (18) leads to the inverse square law.

6.10 Consequence of the General Relativistic Theory

6.10.1 *Relation between metric tensor and gravity*

In accelerative frame, metric tensor g_{44} and g_{k4}, $k = 1, 2, 3$ are related to gravitational scalar potential χ and gravitational vector potential γ_k, $k = 1, 2, 3$. Hence because metric tensor depends on gravitational potential, gravity yields time-space distortion which is defined as derivative of metric tensor with respect to time and space. In accelerative frame, because of time-space distortion, geodesic line is not linear as mentioned above (Paragraph 6.9). The generalized Lorentz contraction (Moller, 1959:244) in case of velocity u, is given by

$$[\{(1 + 2\chi/c^2)^{1/2} - \gamma_k u^k/c\}^2 - u^2/c^2]^{1/2}, \, u^k = dx^k/dt.$$

6.10.2 *Equivalence principle of permanent and artificial gravities*

Einstein insisted equivalence of permanent gravitational field produced by large mass such as Earth and Sun, and non-permanent gravitational field produced artificially such as centrifugal force, by "equivalence principle."

6.10.3 *Relativistic mass*

In general curve coordinate with gravity, mass of particle was expressed by relativistic mass. This depended on gravitational scalar potential χ and gravitational vector potential. If space coordinate is perpendicular to time coordinate axis, gravitational vector potential is zero, and relativistic mass m is expressed by product of mass m_0 at rest, multiplied by the inverse of Lorentz contraction $\{1 + (2\chi - u^2)/c^2\}^{1/2}$ which is generalized Lorentz contraction by considering χ, and velocity u:

$$m = m_0/\{1 + (2\chi - u^2)/c^2\}^{1/2}$$

6.10.4 *Total energy*

The covariant derivative equations of four-dimensional momentum of particle in Minkowski four-dimensional world, consists of four equations. First three equations express motion equations. The fourth equation expresses energy conservation law. The total energy H is given by (Moller, 1959:290)

$$H = m_0 c^2 L(1 + 2\chi/c^2), \, L = 1/[1 + (2\chi - u^2)/c^2]^{1/2} \qquad (6.12)$$

From (6.12), in case of weak gravitational field, neglecting terms of the above square of u/c and leaving till the square of u/c, total energy of particle in stationary gravitational field was given in the following:

$$H = m_0 c^2 + m_0 u^2/2 + m_0 \chi$$

The first term is static energy, the second term is kinetic energy and the third term is gravitational potential energy. In large u, energy could not be separated into kinetic energy and potential energy. The first term is special term in the relativistic theory due to equivalence of mass and energy, where m_0 is the static mass.

6.10.5 *Delay of moving clock*

Advance $\Delta\tau$ of moving clock with velocity u under gravitational scalar potential χ, was given by product of advance Δt of clock in inertial frame multiplied by generalized Lorentz contraction depending on velocity u and χ, if space coordinate is perpendicular to time axis:

$$\Delta\tau = \Delta t\{1 + (2\chi - u^2)/c^2\}^{1/2}$$

In case of clock at rest in the accelerating frame S under gravitational field, $\Delta\tau$ depends only on χ, and Lorentz contraction is $(1 + 2\chi/c^2)^{1/2}$. For example, in case of clock at rest at radius r on disk rotating with angular velocity ω, χ is $- (r\omega)^2/2$, and Lorentz contraction is $\{1 - (r\omega)^2/c^2\}^{1/2}$, and the clock delays as far from center. Observer staying in frame S interprets that the delay of clock is caused by gravitational scalar potential due to centrifugal force in frame S, because clock's velocity being zero in frame S. On the other hand, observer staying in inertial frame I interprets that the delay of the clock at rest in frame S is caused by velocity u of the clock (velocity u is equal to ωr) because there is no gravitational field in inertial frame I (Moller, 1959:245).

6.11 Verification of Correctness of the General Relativistic Theory

6.11.1 *Bending of light*

In stationary gravitational field, light velocity depends on gravitational potential. Because the path of light complies with Fermat's principle (path is determined so as to minimize total time necessary for passing through path), light path is different from linear line due to gravity from equation of light path derived from the Fermat's principle. That is, Einstein concluded that there exists "phenomenon of the bending of light due to gravitation" (Explanation 6.7). He calculated the angle of the bending of light passing near the sun, and predicted that the angle of bending was 1.75″. This prediction could be verified by observation of total solar eclipse without influence of light of the sun. Eddington (Arthur Stanley Eddington)'s observation results is described as mentioned below.

6.11.2 *Perihelion motion of the Mercury*

As phenomenon used to verify correctness of the general relativistic theory, there was "perihelion motion of the Mercury" other than "bending of light due to gravity." This phenomenon could not be explained by Newtonian mechanics. Considering a planet moving in gravitational field of the Sun being much heavy than the Mercury, Einstein estimated that "the perihelion of the Mercury moves due to gravitational field of the Sun," and predicted that the advanced angular change per 100 years was 42.9″, representing the location of the Mercury viewed by the Sun by the angle in revolution plane. This prediction was coincided with the observation results (Explanation 6.8).

Explanation 6.7 Bending of Light in Gravitational Field

We consider propagation of light in gravitational field with spherically symmetric and static matter distribution (Møller, 1959:P. 303). In vacuum, dielectric constant ε and magnetic permeability μ depend on gravitational scalar potential χ as follows: $\varepsilon = \mu = 1/(1 + 2\chi/c^2)^{1/2}$ and then refractive index n given by the square root of product of ε and μ: $n = (\varepsilon\mu)^{1/2}$ depends on χ. Hence, propagation velocity of light w given by light velocity c divided by n, depends on χ: $w = c(1 + 2\chi/c^2)^{1/2}$.

According to Fermat principle, the path of light is decided by solution of Euler equation derived from variational principle minimizing the sum of time $d\sigma/w$ for propagating spatial segment $d\sigma$: $\delta\int d\sigma/w = 0$, $d\sigma^2 = g_{ik}dx^i dx^k$, $i, k = 1, 2, 3$. Euler equation is given by

$$d/d\lambda[g_{ik}(dx^k/d\lambda)] - (1/2)(\partial g_{kj}/\partial x^i)(dx^k/d\lambda)\,(dx^j/d\lambda)$$

$$= -(1/w^3)(\partial w/\partial x^i) \tag{1}$$

$$(d\sigma/d\lambda)w = 1 \text{ (freely selecting parameter } \lambda) \tag{2}$$

In the case of gravitational field as (3) and with polar coordinate in Explanation 6.8, we have

$$w = c(1 - \alpha/r)^{1/2}, \ \alpha/r = -2\chi/c^2 \tag{3}$$

$$(d\sigma/d\lambda)^2 w^2/c^2 = (dr/d\lambda)^2 + [r^2(d\theta/d\lambda)^2$$

$$+ r^2\sin^2\theta(d\phi/d\lambda)^2](1 - \alpha/r) = 1/c^2 \quad \text{(from (2))} \tag{4}$$

For $i = 2, 3$ in (1), we have

$$d/d\lambda[r^2(d\theta/d\lambda)] - (1/2)2r^2\sin\theta \cos\theta(d\phi/d\lambda)^2 = 0 \tag{5}$$

$$d/d\lambda[r^2\sin^2\theta(d\phi/d\lambda)] = 0 \tag{6}$$

The following satisfies (5) and (6),

$$\theta = \pi/2, \qquad r^2(d\phi/d\lambda) = C(\text{constant}) \tag{7}$$

Then (4) becomes as follows:

$$(dr/d\lambda)^2 + r^2(d\phi/d\lambda)^2(1 - \alpha/r)$$

$$= (dr/d\lambda)^2 + C^2/r^2 - C^2\alpha/r^3 = 1/c^2 \tag{8}$$

Introducing $\rho = 1/r$,

$$dr/d\lambda = (dr/d\phi)(d\phi/d\lambda)$$
$$= -(1/\rho^2)(d\rho/d\phi)C\rho^2 = -C(d\rho/d\phi) \tag{9}$$

(8) becomes as follows:

$$(d\rho/d\phi)^2 = 1/\Delta^2 - \rho^2 + \alpha\rho^3 \tag{10}$$

where Δ is defined as $\Delta = cC$.

We consider the light coming from extremely far ($\rho = 0$), with $\phi = 0$. In the case neglecting the term of small α in (10), we have the integral from $\rho = 0$ to $\rho = \rho$:

$$\phi = \int[\Delta/\{1 - (\Delta\rho)^2\}^{1/2}]d\rho = \sin^{-1}(\Delta\rho)$$
$$\rho = (1/\Delta)\sin\phi, \; r = \Delta/\sin\phi \tag{11}$$

Hence in this case, (11) expresses the path of light which passes the point of $r = \Delta$ at $\phi = \pi/2$, and at $\phi = \pi$ go away to extremely far.

On the other hand, in the case of rigorous equation (10), we have the following relation, introducing σ

$$(d\rho/d\phi) = (1/\Delta^2 - \rho^2 + \alpha\rho^3)^{1/2} = (1/\Delta)(1 - \sigma^2)^{1/2} \tag{12}$$
$$\sigma \equiv \Delta\rho(1 - \alpha\rho)^{1/2}, \qquad \text{introducing } \sigma \tag{13}$$

Because of small $\alpha\rho$, neglecting the square of α, we have

$$\sigma \fallingdotseq \Delta\rho(1 - \alpha\rho/2),$$
$$\Delta\rho \fallingdotseq \sigma(1 + \alpha\rho/2) \fallingdotseq \sigma[1 + \alpha\sigma/(2\Delta)] \tag{14}$$
$$\Delta d\rho = d\sigma(1 + \alpha\sigma/\Delta) \tag{15}$$

From (12), we have the integral from $\sigma = 0$ to $\sigma = \sigma$:

$$\phi = \int [\Delta/(1 - \sigma^2)^{1/2}] d\rho$$
$$= \int [(1 + \alpha\sigma/\Delta)/(1 - \sigma^2)^{1/2}] d\sigma$$
$$= \sin^{-1}\sigma - (\alpha/\Delta)(1 - \sigma^2)^{1/2} + \alpha/\Delta \tag{16}$$

The value of $-[\sin^{-1}\sigma - (\alpha/\Delta)(1 - \sigma^2)^{1/2}]$ for $\sigma = 0$ is α/Δ, which is the last term in (16).

The maximum value ρ_m is obtained at $d\rho/d\phi = 0$, hence from (12) and (16), ϕ_m corresponding to ρ_m at $\sigma = 1$:

$$\phi_{m} = \pi/2 + \alpha/\Delta \tag{17}$$

The path of light bends with angle:

$$\Psi = 2(\phi_m - \pi/2) = 2\alpha/\Delta \tag{18}$$

From (14) and $\sigma = 1$,

$$\rho_m = (1/\Delta)[1 + \alpha/(2\Delta)] \tag{19}$$

Since $1/\Delta = \rho_m / [1 + \alpha/(2\Delta)]$, neglecting the square of α,

$$\Psi = 2\alpha\rho_m = 4kM/(c^2 r_m),$$

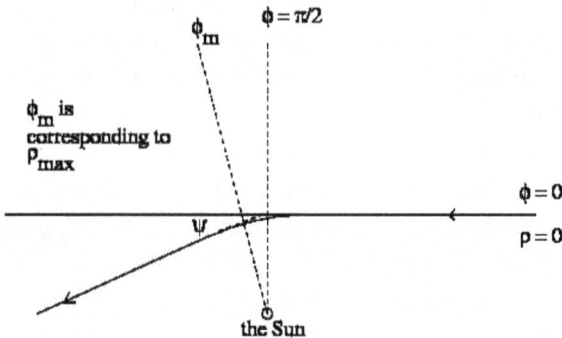

Fig. 6.26. Bending of light passing gravitational field.

r_m: the radius of the Sun, k: Newtonian universal gravitation constant.

Einstein predicted that the light passing through extremely near the Sun bends with bending angle of 1.75″ (Møller, 1959).

6.11.3 *Spectrum shifts*

Because advance of clock depends on gravitational potential, Einstein estimated that "if the difference between gravitational scalar potentials at light emitting place and observing place, is represented as $\Delta\chi$, then spectrum is displaced depending on $\Delta\chi$." That is, when the frequency is ν on observing light with proper frequency ν_0 of atom, and $\Delta\nu$ is defined as difference of frequency $\nu - \nu_0$, then the ratio of $\Delta\nu/\nu_0$ is given by $\Delta\chi/c^2$. He predicted that in case of gravitational field of the Sun, $\Delta\nu/\nu_0$ is -2.12×10^{-6}. When light emitted from atom in surface of the Sun is observed on the Earth, the spectrum of light has slightly small frequency than spectrum of light emitted from atom on the Earth, that is, spectrum shifts to red. Einstein's prediction on spectrum's shift sufficiently coincided with observation results of the Sun and Sirius companion star.

6.11.4. *General relativistic theory guiding cosmic phenomena*

On the general relativistic theory, the verification by experiment is very little. Einstein's general relativistic theory contains Newtonian mechanics as the first approximation. The fact that the difference between Newtonian theory and Einstein's theory appeared in only three phenomena (shift of spectrum due to gravitation, bending of light due to gravitation and perihelion motion of Mercury), indicates that Newtonian theory is a good approximation of gravitational phenomena in the solar system. But on treating cosmic phenomena, it is expected that Einstein's general relativistic theory gives leading guide. For example, theory of black hole which is final stage of star's evolution, is developed on the basis of Schwarzschild solution of gravitational field equation.

6.11.5 *The total solar eclipse observation by Eddington*

Because the general relativistic theory could explain "perihelion motion of the Mercury" which was considered as mystery because of impossibility of theoretical explanation, Eddington was much interested in this theory. He was so enthusiastic commentator that he delivered lecture on the general relativistic theory at meeting of the Royal Society in 1916. In order to verify "phenomenon of bending of light due to gravitation of the sun as Fig. 6.27 predicted by Einstein, in 1919 he leaded members of the total solar eclipse observation to Principe island.

6.11.6 *Bending of light due to gravitation was verified*

On 6th November, at the joint meeting of the Royal Society and the Royal Astronomical Society, from analysis of observation results, the bending

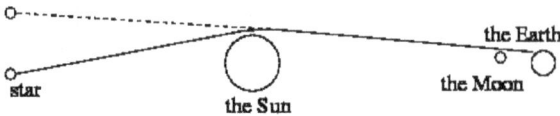

Fig. 6.27. Bending of light passing near the Sun. A star appears in direction of broken line.

Fig. 6.28. Arthur Stanley Eddington (1882–1944).

angle of light passing near the Sun was described as $1.61'' \pm 0.30''$ verifying correctness of $1.75''$ predicted by Einstein. The Times on 7th November with headline of "Revolution in science-new theory of cosmos," introduced Einstein's the general relativistic theory, and reported Eddington's observation results indicating the correctness of the theory. This created a great sensation, and name of Einstein was known in the world.

6.11.7 *Time-space distortion due to gravitation*

When the light passes near the Sun, the light is bended due to passing through time-space distorted by gravitation of the Sun as Fig. 6.29 (Hawking, 2001). The time-space distortion due to gravitation predicted by the general relativistic theory is proved directly by the observation on total solar eclipse by Eddington.

Explanation 6.8 Analysis of the Perihelion Motion of the Mercury

We consider the motion of planet as a mass point in gravitational field. We assume that mass distribution producing gravitational field is static (meaning independence on time), and spherically symmetric. The Sun corresponds to it. We use polar coordinate (r, θ, ϕ) as Fig. 6.30 where the original point is the place of the Sun.

We consider motion of planet with mass m in gravitational field of the Sun with mass *M*. From solution of Newton's theory planet moves on

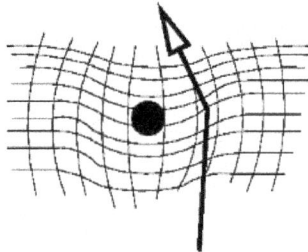

Fig. 6.29. Bending of light passing through time-space distorted due to gravitation of the Sun (Hawking, 2001).

elliptic orbit where distance between planet and the Sun is r, θ is $\pi/2$ (meaning planet being in $x - y$ plane), eccentricity is ε, aphelion r_1 is $d/(1 - \varepsilon)$, perihelion r_2 is $d/(1 + \varepsilon)$, d is given by b^2/a, where semimajor axis a is $(r_1 + r_2)/2$ and semiminor axis b is $(r_1 \times r_2)^{1/2}$.

Eccentricity ε is given by f/a where f is distance between the center of ellipse and focus. In the case of the Mercury, $\varepsilon = 0.2056$, and $d = 5.786 \times 10^{10}$ m.

When aphelion ρ_1 and perihelion ρ_2 are defined as $1/r_1$ and $1/r_2$, respectively, increase of ϕ from aphelion to perihelion is given as follows:

$$\phi_2 - \phi_1 = \pi$$

because $\phi_1 = -\pi/2$ at aphelion ρ_1, and $\phi_2 = \pi/2$ at perihelion ρ_2.

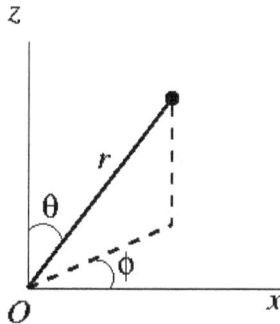

Fig. 6.30. Polar coordinate (r, θ, ϕ).

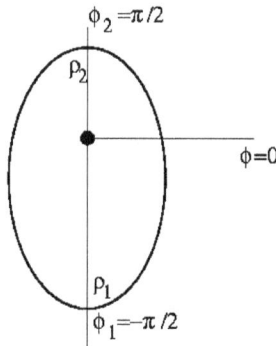

Fig. 6.31. Aphelion ρ_1 and perihelion ρ_2.

Einstein's theory

On the other hand, we intend to obtain $(\phi_2 - \phi_1)$ from equation determining orbit of planet in Einstein's theory. In the case of spherically symmetric and static mass distribution, the solution of gravitational field equation of Einstein is given by Schwarzschild outer solution for outside of matter causing gravity (Møller, 1959:320). Neglecting the term of λ which is important only for large r, the metric tensor in the solution is given by

$$g_{11} = 1/(1 - \alpha/r), \ g_{22} = r^2, \ g_{33} = r^2\sin^2\theta,$$

$$g_{44} = -(1-\alpha/r), \ g_{ik} = 0 \ (i \neq k),$$

$$\alpha = 2kM/c^2 = \kappa Mc^2/(4\pi)$$

$$\chi = -(c^2/2)(1 + g_{44}) = -\alpha c^2/(2r) = -kM/r, \tag{3}$$

where general coordinate is as follow:

$$x^1 = r, \ x^2 = \theta, \ x^3 = \phi, \ x^4 = ct,$$

k, κ and c are defined in Explanation 6.6.

The force received by the planet is

$$\mathbf{K} = (-m\alpha c^2/(2r^2), \ 0, \ 0) \tag{4}$$

where m is the relativistic mass of planet given by

$$m = m_0 L = m_0/\{1 - \alpha/r - u^2/c^2\}^{1/2} \tag{5}$$

$$L^{-1} = \{1 - \alpha/r - u^2/c^2\}^{1/2}: \text{Lorentz contraction} \tag{6}$$

$$u^2 = u_i u^i = g_{il} u^i u^l \tag{7}$$

$$= (dr/dt)^2/(1 - \alpha/r) + r^2(d\theta/dt)^2 + r^2 \sin^2\theta(d\phi/dt)^2$$

Momentum vector $\mathbf{p}$ of planet is given by

$$p_i = mu_i = mg_{ij}u^j$$

$$= m_0 L[(dr/dt)/(1 - \alpha/r), \ r^2(d\theta/dt), \ r^2 \sin^2\theta(d\phi/dt)]$$

Using covariant differential of momentum p_i, the motion equation is given by (Møller, 1959:345)

$$dp_i/dt - (1/2)(\partial \gamma_{kn}/\partial x^i)u^k p^n = K_i \tag{8}$$

$$\gamma_{ik} = g_{ik}, \ p^i = g^{ik}p_k, \ i, k = 1, 2, 3. \tag{9}$$

Equation (8) is expressed as follows:

$$d/dt\{L(dr/dt)/(1 - \alpha/r)\} - (1/2)[-\{(\alpha/r^2)L/(1 - \alpha/r)^2\}(dr/dt)^2$$
$$+ 2rL(d\theta/dt)^2 + 2r\sin^2\theta L(d\phi/dt)^2] = -L[\alpha c^2/(2r^2)], \tag{8a}$$

$$(d/dt)[Lr^2(d\theta/dt)] - r^2\sin\theta\cos\theta L(d\phi/dt)^2 = 0, \tag{8b}$$

$$(d/dt)[Lr^2\sin^2\theta(d\phi/dt)] = 0. \tag{8c}$$

From the energy H of the planet given by (6.12),

$$H/(m_0 c^2) = L(1 + 2\chi/c^2) = L(1 - \alpha/r) = E \tag{10}$$

$$E \text{ is constant. From (8b), } \theta = \pi/2 \tag{11}$$

is a solution of motion equation. From (8c) and (11),

$$Lr^2(d\phi/dt) = \text{constant} = CE \tag{12}$$

C is constant. (10) and (12) leads to

$$r^2(d\phi/dt)/(1 - \alpha/r) = C \tag{13}$$

$$\text{We introduce } \rho = 1/r \tag{14}$$

then from (13), we have

$$dr/dt = (dr/d\phi)(d\phi/dt) = -(d\rho/d\phi)C(1 - \alpha\rho) \tag{15}$$

From (7), (11), (13) and (15), we have

$$u^2 = C^2(1 - \alpha\rho)[(d\rho/d\phi)^2 + \rho^2 - \alpha\rho^3] \tag{16}$$

From (5) and (10),

$$(1 - \alpha\rho)^2 = E^2[1 - \alpha\rho - u^2/c^2] \qquad (17)$$

From (16) and (17),

$$(d\rho/d\phi)^2 = A + B\rho - \rho^2 + \alpha\rho^3 \qquad (18)$$

Denote by ρ_1, ρ_2, ρ_3 the root of $d\rho/d\phi = 0$. Then

$$\rho_1 + \rho_2 + \rho_3 = 1/\alpha \qquad (19)$$

In the Newtonian case, (18) is expressed as

$$(d\rho/d\phi)^2 = A + B\rho - \rho^2 \qquad (20)$$

In this case, we have

$$\phi_2 - \phi_1 = \pi \qquad (21)$$

In the Einstein case (18),

$$d\phi = \pm d\rho/\{(\rho - \rho_1)(\rho_2 - \rho)\}^{1/2}[1 + \alpha(\rho_1 + \rho_2)/2]$$
$$\times [1 + \alpha\rho/2] \qquad (22)$$
$$\phi_2 - \phi_1 = [1 + \alpha(\rho_1 + \rho_2)/2]$$
$$\times \int[(1 + \alpha\rho/2) / \{(\rho - \rho_1)(\rho_2 - \rho)\}^{1/2}] \, d\rho$$
$$= \pi[1 + (3\alpha/4) (\rho_1 + \rho_2)] \qquad (23)$$

where integral $\int$ is carried out from ρ_1 to ρ_2.

We consider the difference $2(\phi_2 - \phi_1)$ of ϕ between two continuous perihelion. The difference $\Delta\phi$ between the result $2(\phi_2 - \phi_1)$ of strict Einstein's theory and the result $2(\phi_2 - \phi_1) = 2\pi$ in Newton's theory, was given by

$$\Delta\phi = 3\pi\alpha(\rho_1 + \rho_2)/2.$$

The result of strict Einstein's theory is larger with $\Delta\phi$ than the result of approximate Newton's theory, and positive $\Delta\phi$ means that every revolution of planet, the perihelion advances. In case of the Mercury, the advance of perihelion predicted by Einstein was 42.9″ per 100 years in angle (Møller, 1959:348). The value coincided with observation result. In case of other planet, advance of perihelion is too small, consequently it cannot be observed certainly.

6.12 The Nobel Prize

6.12.1 *Elsa*

Immediately after living apart from his wife, Einstein found the gravitational field equation (Paragraph 6.8 in this chapter). After living apart from his wife for 5 years, on 14th February 1919 he divorced his wife. As divorce condition, it was written that Mileva previous wife would receive prize money of the Nobel prize which Einstein would be awarded (Pais, 1982:300).

On 2nd June in 1919, he got remarried with Elsa (Elsa Lowenthal) who was cousin known each other since childhood and took care of her cousin during his illness when he suffered from a liver ailment. She was

Fig. 6.32. Einstein and Elsa.

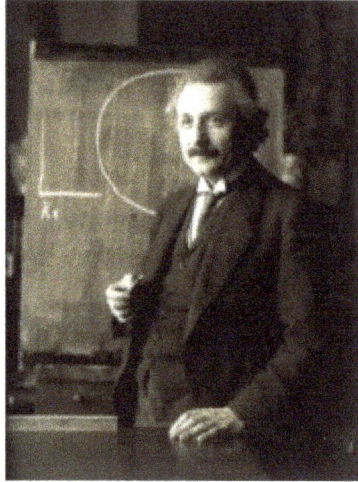

Fig. 6.33.　Einstein in 1921 (photograph in 1921).

born in 1876, and experienced divorce. Her parents lived on lower floors in the same building. They lived at her room after remarriage. Her father was a first cousin of Hermann, Albert's father, and her mother was a sister of Pauline, Albert's mother. Elsa gentle, warm, motherly, and prototypically bourgeois, loved to take care of her Albert. She fell in love with Albert because he played Mozart so beautifully on the violin. Einstein wanted the right to "have his sons visit him in Berlin" to Mileva. To Mileva he sent a note of thanks: I am likewise thankful that you have not alienated me from the children (Isaacson, 2017:217, 228, 229).

In January 1920, Einstein's mother came to Berlin wanting to live with her son. But she was ill with abdominal cancer and passed away in February. In 1918, the First World War ended. The victory allied powers, thrusted severe condition to defeated nation Germany, resulting in unusual economical downfall in Germany. In the year when he lost his mother, seeing the situation of Germany's downfall he recognized that he was German, and he got citizenship of Germany which he gave up at the age of 16, in 1920.

6.12.2 *Visiting the United States and Princeton*

In 1921, he visited the United States for the first time. The main object was to raise funds to create Hebrew University in Jerusalem. For the purpose of the object, at various parts, noisy parade was performed. More than fifteen thousand spectators lined the route, and the crowds cheered wildly (Isaacson, 2017:300). It was rather unusual for a theoretical physicist. They hoped to raise at least $4 million. By the end of the year, only $750,000 had actually been collected (Isaacson, 2017:300). During staying in the United States, he was requested to deliver lectures. After lecture at Princeton, mathematician Veblen (Oswald Veblen) requested the word carved on the stone mantel of the fireplace in the common room in a new mathematics building planned to be completed a decade later, to Einstein. Einstein sent back his approval, and said "Nature hides her secret because of her essential loftiness, but not by means of ruse." Afterward, the building became the home of the Institute of Advanced Study. In 1933, Einstein decided to immigrate to Princeton, and had a office there. He sat in front of the fireplace in the latter part of his life (Isaacson, 2017:297–300).

6.12.3 *Travel abroad for safety*

On 24th June 1922, Einstein's friend Rathenau (Walter Rathenau) who served the Foreign Minister Germany for few months was assassinated by ultra rightist organization Consul. On 4th July Einstein intimate with Rathenau felt danger, and wrote a letter to Curie (Marie Curie) "I should resign a post in Preu*ß*en Science Academy" (Pais, 1982:316). On 8th October, he traveled abroad for safety. On route, Einstein received word that he had been awarded the Nobel prize for his research on photoelectric effect.

The prize money was used for compensation to Mileva previous wife according to divorce condition. In February 1923, he came back to Berlin several months after departure of travel abroad. In this travel, after he stayed shortly in Colombo, Singapore, Hong · Kong and Shanghai, he stayed in Japan for 5 weeks and in Palestine for 12 days. At every place where he went, enthusiastic crowds welcomed him.

6.12.4 *Visiting Japan*

When Yukawa (Hideki Yukawa) was the 4th student of Kyoto the first junior high school of former system, one year before entering the third high school of former system, people in Japan were excited by Einstein's visiting Japan. It was described in autobiography "Traveler — reminiscence of a physicist" that a friend knowing Einstein's news, on physical experiment with partner of Yukawa, said "Ogawa (former name of Yukawa) would become a person like Einstein" (Yukawa, 2011:143). Afterward, Yukawa proceeded to Kyoto University, and in 1935, published paper "Theory of meson" which played important role on interpreting mechanism inside of nucleus, and predicted the existence of meson which was carrier of nuclear force (force between proton and neutron) (Zee, 2010:28). The research result was accomplished with self-supporting ardour in Japan where there was no leader in the field, and was superior to the first class of research in Europe (Nambu, 1998:71). In 1949, he was awarded the Nobel prize for physics due to the research for the first time in Japan. Yukawa would meet with Einstein in the latter part of his life at the Institute of Advanced Study (Paragraph 6.14 in this chapter).

In 1925, Einstein was awarded Copley medal from the Royal Society and in 1926, he was awarded gold medal of the Royal Astronomical Society.

Fig. 6.34.   Hideki Yukawa (1907–1981) (photograph in 1949).

6.12.5 *Einstein and Germany*

Fame attracts envy and hatred. Einstein' was no exception. In this instance, these hostile responses were particularly intensified because of his exposed position in a turbulent environment. During the 1920s, he was a highly visible personality, not for one but for a multitude of reasons.

On 5th May 1916, he succeeded Planck as president of Deutsche Physikalische Gesellschaft. In 1917, he began his duties as director of the Kaiser Wilhelm Institut fur Physik. In 1922, the Akademie appointed him to the board of directors of the astrophysical laboratory in Potsdam. On 12th February 1920, disturbances broke out in the course of a lecture given by Einstein at the University of Berlin.

On 24th August 1920, a newly founded organism, the Arbeitsgemeinschaft deutscher Naturforscher organized a meeting in Berlin's largest concert hall for the purpose of criticizing the content of relativistic theory and the alleged tasteless propaganda made for it by its author (Pais, 1982:312–316).

6.13 The Fifth Solvay Conference

6.13.1 *Discussion on quantum mechanics*

In October 1927, the fifth Solvay conference was held, and discussion was held on quantum mechanics which constituted the two great theories in modern physics with the relativistic theory. Famous scientists developing quantum mechanics attended to the conference. That is, Planck, Bohr (Niels Henrik David Bohr), de Broglie (Louis-Victor Pierre Raymond duc de Broglie), Heisenberg (Werner Karl Heisenberg), Schrodinger (Erwin Rudolf Josef Alexander Schrodinger), Dirac (Paul Adrian Maurice Dirac) and Einstein attended.

6.13.2 *Survey on works of scientists developing quantum mechanics*

To survey works of scientists, developing quantum mechanics is useful to understand the process of foundation of quantum mechanics. First, Planck proposed the concept of energy quantum, and played a role of leading quantum mechanics as mentioned above. Einstein verified the correctness of Planck's concept by theoretical elucidation of photoelectric effect (Paragraph 6.4 in this chapter).

In 1913, Bohr proposed the model of atomic structure. In 1924, de Broglie insisted that all particle had property of wave, and gave the wave length λ by Planck constant h divided by momentum p of particle, that is, $\lambda = h/p$, and he was successful in physical explanation about Bohr's atomic structure model.

Fig. 6.35.   Niels Hendrik Bohr (1885–1962) (photograph in 1922).

Fig. 6.36.   Louis-Victor Pierre Raymond duc de Broglie (1892–1987) (photograph in 1929).

Explanation 6.9 Wave like property of particle

In 1924, de Broglie published the hypothesis:

$$\lambda = h/p \tag{1}$$

The energy of a photon has the energy ε

$$\varepsilon = h\nu \tag{2}$$

which is proposed by Planck, where h is Planck constant and ν is a frequency. de Broglie proposed that the momentum p of the photon is given by

$$p = E/c \tag{3}$$

From (2) and (3), we have the following because of $c = \lambda\nu$,

$$p = h/\lambda \tag{4}$$

This is the relation proposed by de Broglie.

6.13.3 *Wave mechanics and Matrix mechanics*

In 1926, being influenced by wave like property of particle proposed by de Broglie, Schrodinger (Appendix 6.5) derived Schrodinger equation satisfied by wave function representing wave like property of electron, and "wave mechanics" was founded by him. In 1925, one year before then, Heisenberg founded "matrix mechanics" independently of Schrodinger. "Wave mechanics" formulated quantum mechanics with wave function, and on the other hand "matrix mechanics" formulated quantum mechanics with matrix representation. Both theories were different in expression method but were equivalent. Thus, both theories contributed to foundation of quantum mechanics.

6.13.4 *Exclusion principle: fermion and boson*

Pauli (Wolfgang Ernst Pauli) found "exclusion principle." In quantum mechanics, state of electron is represented by discrete state determined by the following: ① energy, ② orbital angular momentum (corresponds to electron's revolution around nucleus), ③ spin (corresponds to electron's rotation). Exclusion principle meant that there could be only one electron in a state of electron.

Particle following exclusion principle such as electron is called as fermion, and particle not following the principle such as photon is called as boson.

Fig. 6.37.   Werner Karl Heisenberg (1901–1976) (photograph in 1927).

Fig. 6.38.   Wolfgang Ernst Pauli (1900–1958) (photograph in 1922).

6.13.5 *Meaning of wave function*

In 1926, Born (Max Born) discovered that the square of absolute value of wave function which is solution of Schrodinger equation, meant probability of presence of particle, giving physical meaning of wave function for the first time. In 1928, Dirac founded theory unifying the relativistic theory and quantum mechanics. He derived Dirac equation concerning spin.

Fig. 6.39.   Max Born (1882–1970).

Fig. 6.40.   Paul Adrian Maurice Dirac (1902–1984) (photograph taken in 1933).

6.13.6 *Probabilistic interpretation in quantum mechanics*

All attendances stayed in the same hotel, and discussed at breakfast and dining. On discussion, Einstein criticized as "probabilistic interpretation in quantum mechanics is uncertain like rolling the dice. Deity does not roll the dice" (Hawking, 2001:37). Bohr considering probabilistic interpretation in quantum mechanics to be appropriate, intensely debated with Einstein. During 4 years, the great debate between Bohr and Einstein was continued. In February 1931, Einstein had accepted Bohr's criticism, and his thought on quantum mechanics changed intensely (Pais, 1982:448). Consequently, in September of the same year, he sent a letter to Nobel committee which described that he nominated Heisenberg and Schrodinger for the Nobel prize.

At the beginning 1928 he suffered from a temporary physical collapse brought on by overexertion. An enlargement of the heart was diagnosed. Hence he had to stay in bed for four months. In 1929, he built small house

Fig. 6.41. Photograph at the 5th Solvay conference.

at Caputh near Berlin. There he had his fiftieth birthday. Several friends presented a sail-boat. For him, sailing on the Havel was one of his fondest pleasures (Pais, 1982:317).

Appendix 6.5 Schrodinger

Schrodinger was born in 1887 in Wien. His father (Rudolf Schrodinger) managed small linoleum factory, and published a paper in the Plants and Animals Society, and acted as vice President of the Society. He was a person with high scholarship and great culture. Because his mother's grand father and mother were Englishman and Englishwoman, English and German were used at home. Because he was educated by private teacher, it was not necessary to go to school till at the age of 11. In 1898, he entered Academy Gymnasium. Here he received education studying Greek and Latin as a priority, and was the top at all school years.

In 1906, he entered Wien University, and majored in physics. He received instruction from Professor Exner (Franz Exner) of experimental physics. 2 years after entrance, he was impressed by listening works on Boltzmann statistics in lecture of Hasenohrl (Friedrich Hasenohrl) who was successor of Boltzmann. He studied methods of mathematical physics, mathematical method treating eigen value problem of partial differential equation. In 1910, he submitted thesis to Wien University and got a doctoral degree.

After service in the army of one year, after graduation, he became assistant of experiment and studied what measure was, viewing directly natural phenomena. His research subject spread to plenty of fields such as electricity, influence of radioactivity on electricity in atmosphere, sound, optics, and color. In 1914, he was qualified as a Professor. The First World War occurred, and he was summoned for army for 4 years, and became artillery at stronghold in west-south front in Austria. But he read scientific papers in the army. In November 1918, after end of war, he came back to the first study room of physics in Wien University. In 1929, he delivered lecture of theoretical physics in Jena University in Germany. Soon afterward, he relocated to Stuttgart University. In 1921, he became Professor in Zurich University. There he research on atomic structure, and was impressed to read paper by de Broglie.

Fig. 6.42. Edwin Rudolf Josef Alexander Schrodinger (1887–1961) (photograph in 1933).

In 1926, he published revolutionary research work on "wave mechanics," and contributed to found quantum mechanics together with Heisenberg as mentioned above. In 1927, he was inaugurated as successor of Planck, head Professor of theoretical physics. There he got acquainted with Einstein. Though he was Catholic, but not Jews, he wished not to stay in Germany which was governed by Nazi oppressing Jews. In 1933, he relocated to Britain due to invitation as Professor at Oxford University. In the same year, he was awarded the Nobel prize of physics. From 1936 to 1938, he served in Graz University, but was dismissed as politically unsuitable person by Nazi. Afterward, he relocated to Dublin, and became President of the Institute of Advance. In 1956, he became special Professor of theoretical physics in Wien University. In 1957, he retired (Hoffmann, 1990).

6.14 Princeton

6.14.1 *The Institute of advanced study*

In 1932, Einstein received the negotiations for invitation to the Institute of Advanced Study. His first plan was to stay in Berlin for 7 months and to stay in Princeton for 5 months. But, this plan did not realized because it was going difficult to stay in Germany due to rise of Nazi. After 3 times

Fig. 6.43. Photograph of Einstein at Princeton in 1935.

meeting with Flexner (Abraham Flexner) the director of the Institute, in October of the same year, his appointment was recognized, and in December he left Germany to the United States. When he locked up his house, he said to his wife that she would never see Caputh again (Pais, 1982:450), because Nazi was much rising in July already.

On 30 January 1933, Hitler (Adolf Hitler) was inaugurated as Prime Minister in Germany. On 28 March, Einstein sent resignation to Preuβen Science Academy. Then Planck who invited Einstein to Preuβen Science Academy, said to his secretary "even if political deep ditch separates me from him, I believe that name of Einstein will be admired as one of the most brilliant star in the Academy for centuries in future" (Isaacson, 2017:406). One week before he sent resignation to Preuβen Science Academy, his house at Caputh was made a raid by German government, but only knife for cutting bread was found (Pais, 1982:450).

6.14.2 *Life Professor at the Institute of Advanced Study*

Hearing news "Hitler was inaugurated as Prime Minister" during staying in Princeton, Einstein decided not to go back to Germany (Raine, 1991:94). In October 1933, he became Life Professor in the Institute of Advanced Study. In 1935, he bought a residence at 112 Mercer Street, and lived there in life. In 1936, his wife suffered from cardiac disease and on

Fig. 6.44.   Otto Harn (1879–1968) (photograph in 1944).

20th December she passed away. Short time after then, he sent to Born a letter in which he said the reason of his non sociability as follows: "I live like a bear in my cave, and really feel more at home than ever before in my eventful life. This bearlike quality has been further enhanced by the death of my woman comrade who was better with other people than I am (Isaacson, 2017:442).

In 1938, Harn (Otto Harn) succeeded nuclear fission reaction in Berlin. In 1939, the Second World War occurred. In this year, Einstein sent a letter informing capability of "construction of atomic bomb" to President Roosevelt as mentioned above, with a warning that German scientists might be pursuing a bomb. In October 1940, Einstein got a citizenship of the United States. In 1941, because intensely working on of Government of Britain, President Roosevelt proceeded in secret "Manhattan Project" for construction of atomic bomb.

In 1944, Einstein resigned the Institute of Advanced Study. After resignation, he continued to research in home going in and out the Institute.

6.14.3 *Interview with Yukawa*

In 1945, atomic bombs were dropped at Hiroshima and Nagasaki, and the Second World War ended. In 1948 after then, he who felt very sad for atomic bomb being used, grasped tightly hand of Yukawa who was staying

Fig. 6.45. The Institute of Advanced Study.

in the Institute of Advanced Study, and said in tears "I apologize to murder Japanese people innocent with atomic bombs" (Yukawa, 1976:200).

6.14.4 *Maja*

His sister Maja (Maja Winteler) and her husband who lived at the small estate outside of Florence bought by Einstein for his sister and husband, were banished by anti-Jewish laws (racial laws) by Mussolini (Benito Mussolini), and her husband (Paul) relocated to Geneva, and sister came to Princeton for living with Einstein. Soon after the end of the war, she began making preparation for rejoining her husband Paul, but the rejoining was not realized. In 1946, she suffered from a stroke and remained bedridden thereafter. Her mind remained clear, but she could no longer speak. Einstein read to her every night after dinner (Pais, 1982:473). The daily work was continued till July 1951 when she passed away.

Because of Einstein's effort to raise funds to create Hebrew university, he served as director of University from 1925 to 1928. In 1948, the doctor diagnosed his illness as abdominal large aorta, one year after then, he was indicated that an aneurysm in the abdominal aorta became larger. In 1950,

Fig. 6.46.   Einstein in 1947.

he began to write his will, and entrusted his scientific documents to Hebrew University. Because in 1952, the first President of Israel passed away, Government requested him to be inaugurated as the second President, but he rejected the request.

Einstein said "The first drop of atomic bomb destroyed not only Hiroshima city but also many. Mankind can be saved only if supranational system is created to eliminate the methods of brute force." One week before he passed away, he wrote a letter to Russell (Bertrand Arthur William Russell). It was a letter agreeing to sign declaration insisting for every nations to abolish nuclear weapons. The desire of Einstein wishing international peace was to abolish nuclear weapons. Now this declaration is called Russell・Einstein declaration.

On 13th April 1955, Einstein collapsed due to rupture of abdominal aorta, but he rejected operation. On 15th, he was carried to Princeton Hospital. In the evening, telephone was made to the elder son (Hans Albert). The elder son hurried to Princeton, and arrived at Hospital afternoon next day. Since 1938, the elder son lived at Berkeley, and since 1947 he served Professor of water mechanics at California University Berkeley school. At quarter past one in the morning on 18th April 1955, Einstein passed away. The funeral was performed by 12 persons containing close friends. After cremation, bone was scattered at the place unknown.

Though Einstein graduated university with superior grades, he could not get a post at anywhere, and spent painful period at beginning of his research life. By the good offices of friend Grossmann he got a post at the Patent Office at Bern. Since then, he continued his research at home utilizing free time after duty. In 1905, his research results were published as three important papers. By these publications, the revolution was introduced to physics. The accumulation of his research at the place different from the academic place like university where he would like to get a position, gave intense impact to academic society. He teaches us that even if being in adversity, it is important to keep enthusiasm, and endeavor for achieving own objective. The relativistic theory by him is called the most beautiful theory made by mankind.

References

Bardeen, J., Cooper, L.N., & Schrieffer, J.R. (1957). Theory of superconductivity. *Physical Review*, **108**(5), 1175–1204.

Baumgardt, C. (1951). *Johannes Kepler: Life and Letters*. New York: Philosophical Library, Inc.

Bowers, B., & Tamura, Y. (trans. 1978), ファラデーと電磁気, Tokyo Tosho, 東京図書. (1974). *Michael Faraday and Electricity*. Wayland Publishers Ltd.

Campbell, L., & Garnett, W. (1882). *James Clerk Maxwell*, London: Macmillan and Co.

Cooper, L.N. (1956). Bound electron pairs in a degenerate fermigas. *Physical Review*, **104**, 1189–1190.

Copernicus, N., & Yajima Y. (trans. 1953), 天体の回転について. Iwanami Shoten, 岩波書店. (1543). On the revolution of the heavenly spheres *[De revolutionibus orbium coelestium]*.

Davy, H. (1816). On the fire-damp of coal mines, and on methods of lighting the mines so as to prevent its explosion. *Philosophical Transactions of the Royal Society*, **106**, 1–22.

Descartes, R., Miyake, T., & Koike, T. (trans. 1993), 方法序説. Hakusuisha, 白水社. (1637). Discours de la method.

Descartes, R., Aoki, Y., & Mizuno, K. (trans. 1993), 屈折光学. Hakusuisha, 白水社. (1637). La dioptrique.

Descartes, R., & Akagi, S. (trans. 1993), 気象学. Hakusuisha, 白水社. (1637). Les meteors.

Descartes, R., & Hara, R. (trans. 2013), 幾何学. Tikuma Shobo, 筑摩書房. (1637). La geometrie.

Drake, S., & Akagi, A. (trans. 1993), ガリレオの思考をたどる. Sangyo Tosho. 産業図書. (1990). *Galileo: Pioneer Scientist*. University of Toronto Press.

Faraday, M. (1832). Experimental researches in electricity, *Philosophical Transactions of the Royal Society*, **122**, 125–162.

Faraday, M., & Takeuchi, Y. (trans. 2010). ロウソクの科学, Iwanami Shoten, 岩波書店. (1861). The chemical history of a candle.

Galilei, G., Konno, T., & Hida, S. (trans. 1937–1948), .新科学対話上・下. Iwanami Shoten, 岩波書店. (1638). Dialogue concerning two new sciences.

Galilei, G., & Aoki, Y. (trans. 1959–1961), 天文対話 上・下, Iwanami Shoten, 岩波書店. (1632). Dialogue concerning the two chief world systems.

Galilei, G., Yamada, K., & Tani, Y. (trans. 1976), 星界の報告. Iwanami Shoten, 岩波書店. (1610). Sidereus nuncius.

Gleick, J. (2003). *Isaac Newton. London: Forth Estate*. A division of Harper Collins Publishers.

Hawking, S., & Sato, K., (trans. 2001). ホーキング、未来を語る. Artist House, アーティストハウス. (2001). *The Universe in a Nutshell*. The book Laboratory Inc.

History, C. (2019). Galileo Galilei: A captivating guide to an Italian astronomer, physicist and engineer and his impact on the history of science. *Captivating History*.

History, H. (2019). *James Clerk Maxwell: A Life from Beginning to End*. Hourly History.

Hoffmann, D., & Sakurayama, Y. (trans. 1990). シュレーディンガーの生涯. Tijinshokan, 地人書館. (1984). Erwin Schrodinger. Verlagsgesellschaft. Leipzig.

Isaacson, W. (2017). *Einstein: His Life and Universe*. London: Simon & Shuster UK Ltd.

James, F.A.J.L. (Ed.). (1991). *The Correspondence of Michael Faraday*, Volume 1, pp. 1811–1831. London: The Institution of Electrical Engineers.

James, F.A.J.L. (2010). *Michael Faraday. A Very Short Introduction*. Oxford University Press.

Kekule, F.A. (1865). Sur la constitution des substances aromatiques. *Bulletin de la Societe Chemique de Paris*, **3**, 98–110.

Kepler, J., & Kishimoto, Y. (trans. 2009). 宇宙の調和―不朽のコスモロジー. Kosakusha, 工作舎. (1619.). *The Harmony of the World* [Harmonices mundi libri]. Lincii Auftriae.

Kepler, J., & Kishimoto, Y. (trans. 2013). 新天文学―楕円軌道の発見. Kosakusha. 工作舎. (1609). *Astronomia Nova*.

Kepler, J. (1596). *Mysterium Cosmographicum. (The Cosmographic Mystery).*

Maxwell, J.C. (1855–56). On Faraday's lines of force. *The Transactions of the Cambridge Philosophical Society*, **X**, Part 1.

Maxwell, J.C. (1861–62). On physical lines of force. *The Philosophical Magazine*, **XXI**.

Meissner, W., & Ochsenfeld, R. (1933). Ein neuer Effekt bei Eintritt der Supraleit fahigkeit. *Die Natur-wissenschaften*, *21*, 787–788.

Mendelssohn, K., & Ooshima, K. (trans. 1971). 絶対零度への挑戦. Kodansha. 講談社. (1966). *The Quest for Absolute Zero*. Weidenfeld and Nicolson.

Moller, C., Nagata, T., & Ito, D. (trans. 1959). 相対性理論. Misuzu Shobo. みすず書房. (1952). *The Theory of Relativity*. Oxford University Press.

Michelson, A., & Morley, E. (1887). On the relative motion of the Earth and the luminiferous ether. *American Journal of Science*, **34**(203), 333–345,

Nambu, Y., 南部陽一郎. (1998). クォーク 第二販 (*Quark*, 2nd edn.). Kodansha. 講談社.

Newton, I., & Nakano, S. (trans. 1977). プリンキピアー自然哲学の数学的原理. Kodansha. 講談社. (1726). *Philosophiae Naturalis Principia Mathematica*, 3rd edn.

Newton, I., & Shimao, N. (trans. 1983). 光学. Iwanami Shoten, 岩波書店. (1721). *Optics*, 3rd edn.

Pais, A. (1982). Subtle is the lord. *The Science and the Life of Albert Einstein*. Oxford: Oxford Univ. Press.

Raine, D. J., & Okabe, T. (trans. 1991). アインシュタインと相対性理論. Tamagawa University press, 玉川大学出版部. (1975). *Albert Einstein and Relativity*. Wayland Publishers Ltd.

Shimao, Y. 島尾永康 (2000). ファラデー ＝ 王立研究所と孤独な科学者 (*Faraday=Royal Institution and Alone Scientist*). Iwanami Shoten, 岩波書店.

Shioyama, T. 塩山忠義 (2002). センサの原理と応用 (*Principle of Sensor and its Application*), Morikita Shuppan, 森北出版.

Sootin, H., Koide, S., & Tamura, Y. (trans. 1976), ファラデーの生涯. Tokyo Tosho, 東京図書. (1954). *Life of Michael Faraday*. New York: A division of Simon & Schuster Inc.

Sootin, H., & Watanabe, M. (supervised the translation), Tamura, Y. (trans. 1977), ニュートンの生涯, Tokyo Tosho, 東京図書. (1955). *Isaac Newton*. London: Julian Messner. A division of Simon & Schuster, Inc.

Sugget, M., & Oohashi, K. (trans. 1992), ガリレオと近代科学の誕生. Tamagawa University Press, 玉川大学出版部. (1981). *Galileo and the Birth of Modern Science*. Wayland Publishers Ltd.

Tomonaga, S. 朝永振一郎 (1958). 量子力学 *(Quantum Mechanics)*, Misuzu Shobo, みすず書房.

Tyndall, J. (2002). *Faraday as a Discoverer*. McLean, VA: IndyPublish. com. (Original work published in 1868, New York: D. Appleton).

Voelkel, J.R., & Hayashi, D. (trans. 2010). ヨハネス・ケプラー：天文学の新たなる地平へ, Ootuki Shoten, 大月書店. (1999). Johannes Kepler. Oxford University Press.

White, M. (1998). *Isaac Newton: The Last Sorcerer. London: Fourth Estate.* A division of Harper Collins Publishers.

Yukawa, H. 湯川秀樹 (1958). 旅人: ある物理学者の回想 *(Traveler.: Reminiscence of a Physicist)*. Asahi Shinbunsha, 朝日新聞社.

Yukawa, H., & Tamura, S. 湯川秀樹, 田村松平 (1955–1962). *Accepted Theory of Physics* (Vols. I–III). 物理学通論 上中下, Tokyo: Taimeido. 大明堂.

Yukawa, S. 湯川スミ (1976). 苦楽の園 *(Garden of Pleasure and Pain)*. Kodansha, 講談社.

Zee, A. (2010). *Quantum Field Theory in a Nutshell*. Princeton: Princeton University Press.

Chronological Table

	Galileo
1564	On 15th February Galileo Galilei was born in Pisa
1581	Enrolled at the University of Pisa
1585	Left halfway the university of Pisa without completing his degree
1586	His first paper "The little balance"
1589	Professor of Mathematics in the University of Pisa
1591	His father Vincenzio passed away
1592	Professor in the University of Padua. Research on falling body
1597	Sent the letter for Kepler saying his support of heliocentric system
1599	Marriage with Marina Gamba
1601	Private teacher for Cosimo II, the son of Ferdinand I
1608	Cosimo II became the Grand Duke of Tuscany
1609	Fabricated a telescope himself. Observed the Moon.
1610	Discovered satellites of Jupiter. Published "Report of Stars."
1611	Member of Accademia Lincei
1613	Two daughters enrolled in St. Matteo Monastery
1616	The first trial in the Inquisition
1632	Published *Dialogue Concerning the Two Chief Systems*
1633	The second trial in the Inquisition • Condemned to lifelong imprisonment • Lived with the Archbishop of Siena • Returned to his home in Arcetri

1634	Maria Celeste passed away
1637	Lost one's sight
	Published *Discourse and Mathematical Demonstrations Concerning the Two New Systems*
1642	Passed away in Arcetri

Kepler	
1571	On 27th December Johannes Kepler was born in Weil der Stadt
1577	Great Comet appeared
1584	Enrolled in the elementary seminary in Adelberg
1586	Enrolled in the advanced seminary, Seminare Maulbronn
1587	Enrolled in the University of Tubingen
1588	Bachelor degree in the University of Tubingen
1589	Registed Stift in the University of Tubingen
	Learned Mathematics and Astronomy from Michael Maestlin
1591	Master degree
1594	Taught Mathematics and Astronomy at Graz's seminary
1596	Published *Mysterium Cosmographicum*
1597	Got married to Barbara Muller
1599	Assistant of Tycho Brahe in Praha
1601	Father-in-law passed away
	Tycho Brahe passed away
	Kepler was appointed as the Mathematician for Rudolf II
1609	Published *Astronomia Nova*
1612	Rudolf II passed away
1618	Defenestration of Praha
1619	Published *Harmonice Mundi*
1621	In witch trial, the court ruled that his mother was not guilty
1626	A big army of farmers surrounded Linz and set fire to the edge of the city. Printer was burned. Kepler's family left for Regensburg
1627	Completed "Rudolfphine table"
1630	Visited an autumn market in Leipzig
	On 15th November he passed away

Newton	
1643	On 4th January Isaac Newton was born in Woolsthorpe
1646	Mother Hannah's remarriage with Barnabas Smith
1649	Charles I was headed. Puritan revolution started
1653	Cromwell became Lord Protector
1653	Step father Smith passed away
1654	Entered King's school
1658 ∫ 1660	Absence from King's school
1660	Charles II took the throne. Restoration occurred
1661	In June, entered Trinity College of Cambridge University
1663	Isaac Barrow was inaugurated as the first Lucasian Professor of Mathematics
1665	Got a Bachelor of Arts
1665 ∫ 1667	University was closed due to the plague. Research at home in Woolsthorpe
1667	Became a Minor Fellow
1668	Got Master of Arts
1669	Was inaugurated as Lucasian Professor of Mathematics
1671	Presented his own reflecting telescope to the Royal Society
1672	The first paper "Theory of Light and Colors" was published in the *Philosophical Transactions of the Royal Society*
1677	Isaac Barrow passed away
1679	Mother Hannah passed away
1684	Edmond Halley visited Cambridge, and requested manuscript to Newton
1685	James II took the throne
1687	*Principia* was published
1689	William took the throne
1696	Was inaugurated as Warden of the Mint
1699	Was inaugurated as Master of the Mint
1701	Resigned Professor of Cambridge University
1703	Was inaugurated as President of the Royal Society
1705	Was Knighted
1727	On 20 March passed away in Kensington, buried in Westminster Abbey

Faraday	
1791	On 22 September Michael Faraday was born at Butts at the edge of London
1793	Louis XVI and Queen Marie-Antoinette were executed
1800	Alessandro Volta invented battery
1804	was employed as a newspaper-cumerrand boy at bookshop and stationers. Napoleon became Emperor
1805	became apprentice at the bookshop
1810	attended the lecture by John Tatum
1812	attended the lecture by Humphry Davy. Completed apprenticeship, relocated to DeLa Roche's shop
1813	In January met with Humphry Davy 1 March served as experimental assistant in the Royal Institution In October departed to Continent with Humphry Davy
1815	In April ended journey of Continent, arrived at London
1820	Oersted's discovery
1821	Success in reverse problem of Oersted. Marriage with Salah
1823	Success in liquefaction of chlorine gas
1824	Elected as Fellow of the Royal Society
1825	Became head of laboratory in the Royal Institution. Discovered Benzene. William Sturgeon invented electromagnet
1826	Started the Christmas Lecture
1829	Humphry Davy passed away in Geneva
1829	Inaugurated as Professor of Chemistry in English Military Academy
1831	Submitted paper of electro-magnetic induction. Published in 1832
1833	Discovered laws of electro-chemical decomposition
1836	Inaugurated as adviser of Trinity House
1840	Invented new type chimney of oil lamp
1844	Requested to try the explosion accident in Haswell Colliery from Prime Minister Peel
1845	Discovered Faraday effect. Discovered diamagnetism
1853	Crimean War (~ 1856)
1854	Opposition to use chemical weapon
1858	Was offered to use a grace and favor house on Hampton Court from Queen Victoria
1867	On 25 August passed away at Hampton Court. Buried at High gate Cemetery

Maxwell	
1831	On 13 June, James Clerk Maxwell was born in Edinburugh
1837	The woodcut expressed 6 years old James looking at a bow of violin
1839	On 6 December, mother Frances passed away
1841	Enrolled in the Edinburgh Academy
1842	In February, father planned to see electro-magnetic machines
1844	13 years old James fabricated five regular polyhedrons
1846	On 6 April, James' paper on oval curve was published in Proceedings of the Edinburgh Royal Society, vol. ii, pp. 89–93
1847	Enrolled in the University of Edinburgh
1850	In April, migrated in Peterhouse College of Cambridge
1851	In April or May, Experiment of Foucault's pendulum in Trinity College
1852	In April, got scholarship
1853	The second wrangley in Cambridge
1855 –6	The paper "On Faraday's Lines of Force."
1856	On 2 April, father Mr. John Clerk Maxwell passed away In November, Perofessor of Natural Philosophy in Marischal College at Aberdeen
1858	Got married to Katherine Mary Dewar
1860	In summer, Professor of Natural Philosophy in King's College, London
1861	On 17 May, at the first time, lectured "On the Theory of the three Colours"
1865	Resigned the Professor in King's College, returned to Glenlair.
1866	The paper "On the Viscosity of Gases" published in the *Philosophical Transactions*
1871	In March, the Chair of Cavendish Laboratory in the University of Cambridge
1876 –77	In November, a member of the Council of the Senate of the University of Cambridge. The president of the Cambridge Philosophical Society.
1879	On 5 November, passed away

Einstein	
1879	On 14 March Albert Einstein was born in Ulm
1888	In October entered Luipold Gymnasium
1895	Failure in entrance examination of ETH
1896	In Oct entered ETH
1900	Qualified as a Fachlehrer. Max Planck proposed the idea "energy quantum"
1902	Employed temporally in the Patent Office in Bern. In 1904, served as a regular staff
1903	Marriage with Mileva
1904	Hendrik Lorentz derived Lorentz transformation
1905	Published papers "photoelectric effect," "the special relativistic theory," "Brownian motion"
1906	Ludwig Boltzmann passed away
1907	Solved problem in specific heat of solid
1908	Got a post of private lecturer
1909	Got a post of Associate Professor at Zurich University
1911	Inaugurated as Professor of the Karl-Ferdinand University in Prague
1912	Inaugurated as Professor of ETH Zurich
1913	Published paper of the general relativistic theory with Marcel Grossmann. Niels Bohr proposed atomic structure model
1914	Inaugurated as Professor of Berlin University. The First World War occurred
1916	Published complete version of the general relativistic theory
1919	Arthur Eddington verified Einstein's prediction on "phenomenon of bending of light due to gravitation of the sun" by total solar eclipse observation. Divorce with Mileva. Remarriage with Elsa
1921	Visited the United States to raise funds to create Hebrew University
1922	Awarded Nobel prize of Physics
1924	De Broglie proposed wave like property of particle and de Broglie's relation between wave length and momentum
1925	Werner Heisenberg founded "matrix mechanics"

1926	Erwin Schrodinger founded "wave mechanics"
1932	John Cockcroft and Ernest Walton at Cavendish Institute verified "mass-energy equivalence" by nuclear fission reaction
1933	Hitler was inaugurated as Prime Minister
	Inaugurated as Life Professor in the Institute of Advanced Study
1936	Elsa passed away
1938	Otto Harn succeeded nuclear fission reaction
1939	The Second World War occurred. Sent a letter notifying "possibility of construction of atomic bomb to President Roosevelt
1944	Resigned the Institute of Advanced Study
1945	Atomic bombs were dropped in Japan. The Second World War ended
1952	Rejected request to be the second President of Israel
1955	On 18 April passed away in Princeton. After cremation bone was scattered by his son and friends
	Published *Russell–Einstein Declaration*

Name Index

Subject Index

About the Author

Tadayoshi Shioyama (塩山忠義) has received his bachelor's degree in Physics in 1966 and doctorate degree in Engineering at the Department of Mathematics and Applied Physics in 1984 from Kyoto University.

He has published the following books: Shioyama, T. (2002). *Principle and Application of Sensor* センサの原理と応用. Morikita Shuppan, 森北出版; Shioyama, T. (2010). *Basic and Application of Image Understanding and Pattern Recognition*, 画像理解・パターン認識の基礎と応用, Trikepus, トリケップス; Shioyama, T. (2019). *Newton-Faraday-Einstein*, ニュートン・ファラデー・アインシュタイン, Nakanishiya Shuppan, ナカニシヤ出版; and Shioyama, T. (2021). *Newton・Faraday・Einstein*, World Scientific Publishing Co. Ltd.

www.ingramcontent.com/pod-product-compliance
Lightning Source LLC
Chambersburg PA
CBHW050543190326
41458CB00007B/1897